TECHNOLOGICAL ISSUES IN BROADCAST EDUCATION

Critical Challenges

Edited by Gerard Donnelly and Joseph R. Blaney

PRAEGER

Westport, Connecticut
London

Library of Congress Cataloging-in-Publication Data

Technological issues in broadcast education : critical challenges / edited by
Gerard Donnelly and Joseph R. Blaney.
p. cm.
Includes bibliographic references and index.
ISBN 0-275-97521-5 (alk. paper)
1. Broadcasting—Study and teaching. 2. Broadcasting—Technological innovations. 3. Broadcasting. I. Donnelly, Gerard. II. Blaney, Joseph R.
PN1990.83.T43 2003
384.54'071—dc21 2003051776

British Library Cataloguing in Publication Data is available.

Library of Congress Catalog Card Number: 2003051776
ISBN: 0-275-97521-5

First published in 2003

Praeger Publishers, 88 Post Road West, Westport, CT 06881
An imprint of Greenwood Publishing Group, Inc.
www.praeger.com

Printed in the United States of America

The paper used in this book complies with the Permanent Paper Standard issued by the National Information Standards Organization (Z39.48–1984).

10 9 8 7 6 5 4 3 2 1

Copyright Acknowledgments

The author and publisher gratefully acknowledge permission to use excerpts from the following material:

Interviews and personal communications with Michael Ameigh, Alan Albarran, Sylvia Chan-Olmsted, Roger Cooper, William Covington, Jr., Vince Doody, Susan Tyler Eastman, Douglas Ferguson, Barbara Croll Fought, Lionel Grady, Ron Graeff, Lynne Gross, Linwood Hagin, Phillip Heeler, Stephen Jacobs, Karen Kearns, Michael Keith, Norm Medoff, Robert Musberger, Peter Orlik, Robert Papper, Larry Patrick, Sam Sauls, Lynn Spangler, Carol Spradling, Jodell Strauch, Tom Streeter, Judith Thorpe, Doug Underwood, Craig Warner, Randall Wenner, and Laura Widmer.

To Kathy, Brian, Erin, and Lauren, with love
And to my parents, Catherine and John Donnelly,
for life's lessons well taught
—Gerard Donnelly

To Lauri, Maggie, Matthew, and Clare
—Joseph R. Blaney

Contents

Acknowledgments

A collection of essays of this scope required contributions from many sources. The editors wish to thank each of the authors as well as the Broadcast Education Association, which provided opportunities to meet these contributors. Carolyn Toomey of Argosy Publishing was most helpful at the production stage of this book. Likewise, we are thankful for the myriad project and acquisition editors from Greenwood Publishing.

In addition, we would like to express our appreciation to our university colleagues: Dr. Ron C. DeYoung, Dean of the Melvin and Valorie Booth College of Business and Professional Studies, Dr. Taylor Barnes and Dr. Tim Gilmour of the Office of the Provost, and Ms. Marla McCrary of the Department of Mass Communication, all at Northwest Missouri State University, for their financial and administrative support of this project; Dr. Roberta Trites, Dean of the College of Arts and Sciences, and Dr. Larry Long, Chair of the Department of Communication, both at Illinois State University, for their moral support.

We would like to thank Sybil Brianne Wyatt for her research assistance.

Finally, we wish to thank our respective spouses, Kathy Donnelly and Lauri Blaney, for their love and professional support.

Introduction

This book literally started out over coffee. At the April 1999 Broadcast Education Association meeting, coeditors Donnelly and Blaney had gathered together a number of BEA faculty to participate in a discussion of how technological considerations were impacting their mission. Data collected at this meeting and surveys of 234 BEA institutional administrators resulted in a subsequent article (Blaney and Donnelly, 2000) about relationships between educational broadcast facilities and professional preparation, enrollment, alumni satisfaction, and faculty satisfaction. While informative, this study raised more questions in need of answers. As such, the faculty gathered in Las Vegas once again in 2000 and decided that an edited volume would serve the purpose of answering the disparate questions that arose from these conversations.

This book is a compilation of answers to those diverse questions. The chapters have one unifying purpose. They each seek to illuminate the educational settings that have been created and/or enhanced by the emergence of new broadcast-related technologies. In Chapter 1, W.A. Kelly Huff explains how the word "digital" has been incorporated into the broadcast educator's lexicon.

In Chapter 2, Matt Jenkins addresses the question of how the introduction of new technologies in broadcast education has created choices about how to teach certain skills in the changing technological environment. Moreover, he asks whether we are truly teaching anything new.

In a chapter sure to be of interest to administrators, Steve Craig discusses the perennial question of funding. Chapter 3 also offers the intriguing story of the use of student fees at the University of North Texas.

In Chapter 4, Jerry Condra addresses the boundaries of broadcast course work that new technologies have breached. Taking a larger snapshot, Gerard Donnelly asks similar questions about boundaries between academic departments in Chapter 5. In Chapter 6, Doug Sudhoff and Rick

Musser trudge through the pedagogical muck of professional and academic cultures to discover more effective ways of teaching convergent reporting.

Chapters 7 and 8 strike at the heart of classroom (and nonclassroom) behaviors. Joyce Chen examines the impact of new technologies on the teaching-learning relationship. Then, Joseph Blaney, Gerard Donnelly, and Matt Rouch ask whether differences exist between traditional and modular broadcast production course populations.

In Chapter 9's provocative essay, Dom Caristi raises legal questions about the digital nature of the broadcast classroom and laboratory. Perhaps just as provocative, Mark Braun and Rebecca Lind address race and gender as they discuss the pedagogical use and evaluation of technologies. Chapter 10 also has implications for our understanding of the "digital divide."

Bruce Mims offers wise counsel to broadcast educators who may be undertaking digital conversions. In this sense, Chapter 11 will be of interest to broadcast lab engineers as well as instructors.

Kevin Lee addresses the increasingly common expectation that much student learning will take place outside the classroom. He evaluates the potential for learning from materials in different extra-classroom settings.

Finally, Jong Kang and Seok Kang offer a comparative analysis of broadcasting curriculum from three different national models. The implications of technological developments are inseparable from the discussion of these differing approaches.

In short, this book casts a wide net and offers up diverse thoughts about how technology is impacting broadcast education at the beginning of the 21st century. The topics addressed are as varied as the methodological approaches employed. Nonetheless, one thread runs through all the chapters: the idea that the capabilities of new technologies in broadcast education are providing solutions . . . . and critical challenges as well.

We hope you enjoy this collection.

CHAPTER 1

New Radio-TV Technologies and "The Digital Perception"

W.A. Kelly Huff

New terminology and abbreviations have made understanding electronic media more difficult than ever. Confusion reigns among industry professionals and consumers alike. There was a time when two of the least complicated household appliances were the family television and radio set. With either one, the consumer needed only to turn it on and use it. Although that is still the case theoretically, it is widely recognized that there is much more to using electronic media equipment these days than merely switching on the power. The integration of computer technology with home audio and video electronics, creating digital media, has made the learning curve much greater for operation. Computers were created to help simplify our lives, but Beacham (1997b) warned: "Despite claims to the contrary, these are not yet simple appliances average people can easily set up and productively use" (p. 24).

Radio and TV technologies are emerging in such rapid fashion that even professionals, educators, and students struggle to keep up with the pace. Many who fall into one of those categories are typically ahead of the learning curve, so pity the average consumer who is trying to make sense of it all. Retailers are of minimal help, because few sales personnel have any in-depth understanding of the items they are selling. Anyone wanting to cloud an issue, to promote a product, or to claim expertise can spit out a slew of letters, numbers, acronyms, and names (see Glossary). Further, electronic equipment is frequently "set up improperly in stores, especially electronics superstores," so consumers should be especially careful in making "any judgments based solely on that kind of experience" (Feedback, 2001, p. 18). It has reached the point where one needs a scorecard to keep up with it all—and that probably would not help much. Like it or not, new terminology, acronyms, and abbreviations are part of modern society and they appear to be here to stay.

Much of the present confusion seems to have been caused by the introduction of "digital" technology. More often than not, the word "digital" is abused by those who either do not know what it is, or by those who use it as the promotional buzzword that it has become. In fact, the aforementioned retailers are probably the biggest abusers of the word "digital," which "clearly has been positioned in our modern technological language to define the zenith of audio" and video (Hahn, 1994, p. 27). Consumers accept the fact that they will "pay a premium" if the electronic media component is "digital," even if it is not "any better than other versions previously available" (p. 27). Genuine digital technology does offer benefits over analog; however, as a term it is almost universally misused, abused, and overused (Nelson, 1993, p. 94). To consumers, a "digital" product conjures up the notion of an electronic component that "automatically offers some kind of superior quality" (p. 94). By comparison, older electronic devices are inferior to these "smarter, faster, cheaper digital" ones (p. 94). As Pohlmann (1992) observed: "No one can resist the promise of future technology" (p. 62).

This digital revolution might represent the desire of consumers to have something different. Hirsch (1992) observed: "A never-ending drive for improvement of our lives seems to be an intrinsic part of human nature" (p. 26). Such a concept is not new, particularly in the world of media and electronics. Some consumers are "future junkies," who are always "clamoring for next year's products," and they live their lives in a state of anxiety that their "cherished gear will grow obsolete" (Pohlmann, 1992, p. 62). With digital, perception may truly have become a reality. Rare is the advertisement for electronic equipment, particularly audio or video, that does not boast products that are digitally encoded, enhanced, processed, synthesized, altered, or remastered. Ken Hahn (1994), a three-time Emmy-award winner for audio, admitted: "I'm 'in the business' and I don't know what some of these things really are or mean or how they might improve or affect performance. I do know they affect price" (p. 27). Thus, one must question how the average consumer of audiovisual components can understand the difference between analog and digital equipment. Like analog, digital quality exists in varying degrees, running the gamut from "high-end" to "low-end" (p. 54). Hahn added: "Is the digital version any better than other versions previously available? Whether or not it is true, it is perceived as such. . . . However, let's not allow the cloud of confusion that envelops consumer 'digital' audio electronics to pollute our professional environment" (p. 27).

DIGITAL BROADCASTING

We are in the midst of a transition to digital radio and television broadcasting in the United States, but there are those who question the necessity of converting from our analog transmission systems. Haber (1996) noted: "An industry observer who wished to remain anonymous questioned whether people who listen to radio could tell the difference 'between a radio station that's doing digital broadcasting and a radio station that's broadcasting in analog.' The source said, 'We keep forgetting that John Q. Public is a non-educated, non-informed listener. Most people couldn't tell you that the radio station they're listening to plays compact discs or (is) running analog tape'" (p. 63).

Does it really matter if audio—broadcast or otherwise—is digital, especially given that the perception that it is digital may be enough to satisfy consumers? Diamond and Sneegas (1992), two researchers investigating the impact of labeling and image on the selection of audio, conducted a comparison test to determine the extent of perception when it comes to audio. They packaged a single audio product under three different labels. In their study, they let the consumer see the label. The researchers discovered that audio labeled AM consistently evoked the most negative listener reaction. Using these results, it might actually be better to rename AM and FM analog broadcasting systems rather than change them and use the same name. Indeed, that is what Canada is considering as the country's communication officials have decided to employ terrestrial digital audio broadcasting (DAB) by using IBOC (in band, on channel) DAB over L-band rather than AM or FM. The strategy might have merit, but the United States will probably institute IBOC using the same spectrum currently devoted to AM and FM analog broadcasting. Will American consumers hear a digital difference in audio that is broadcast on the traditional AM and FM frequencies? If Diamond and Sneegas's (1992) findings are correct, then the answer is probably not. They concluded: "The central issue is not whether we can or cannot, or whether we should or should not, but whether the consumer wants, and will purchase, the new technology. The cost to broadcasters to compete in the arena of emerging technologies is high. The cost not to compete is even higher" (p. 20).

Although much of its history is devoid of profound and dramatic technological advances, radio has evolved when operators found it absolutely necessary to do so. Both radio and TV were forced to evolve to remain viable in the marketplace with emerging technologies. FM stereo was implemented in 1961 not only to broadcast stereo recordings but also to enable FM to compete with AM. Likewise, AM stereo was conceived in the 1970s to help AM compete with FM stereo. DAB, similarly to FM stereo, was motivated by the emergence of digital compact discs, or CDs, in 1981. DAB is considered CD-quality, which is another new buzzword that has replaced FM-quality as the measure for audio excellence. Although an overwhelming number of stations have been equipped with digital production equipment for many years, listeners were not impacted because the signal was delivered to them in analog form. There has been no way to hear the full impact of CDs on analog broadcasting stations. Consumers became interested in color TV due to the appeal of Technicolor movies. Multichannel television sound (MTS) and digital television (DTV) were instituted because of videocassette recorders (VCR) and videodisc players. Both VCRs and videodisc players offered dramatic audio improvements over traditional monophonic broadcasting. Digital was brought to the forefront of consumers' minds because of the home computer market. Forrest (1994) observed: "[Since then,] there has been a relentless move toward digital audio, radio, and television. CDs were the most obvious manifestation of digital technology as far as the public was concerned, bringing a real benefit in sound quality and durability of the recording" (p. 5). Digital versatile discs (DVD) have served as a similar catalyst in the video market.

What does "digital" mean?

With all the hoopla surrounding it, it is of course prudent to understand what digital is. Digital is a computer language in which information is converted to zeros and ones, which are otherwise known as binary digits or bits. These bits are the smallest pieces of video or sound information in a digital system. Upon conversion, the information "can be fed to and stored in computers as long strings of bits," known as "digital information" (Gates, Myhrvold, & Rinearson, 1995, pp. 23-24).

So, does that make digital delivery better than analog? In radio and television broadcasting terms, the analog signal is "a continuously variable and varying signal" (Mirabito, 1994, p. 26). A digital signal "is not a continuously variable and varying signal" and "assumes only a finite number of discrete values" (p. 27). The primary advantage of digital technology over analog is that more television or radio channels (or information) can be squeezed into the same spectrum space presently occupied by analog-delivered signals. Information delivered digitally is virtually unlimited, and a television set may be operated more like a multimedia computer (Doyle, 1992). Mannes (1991) elaborated: "An uncompressed, digitized HDTV [high-definition television] signal might require more than *a billion* bits per second," and "storing those billion bits would require more than 80 $3\frac{1}{2}$-inch floppy disks" (p. 30). Digital technology allows that tremendous amount of information to be compacted, or compressed.

DIGITAL COMPRESSION

Compression, otherwise known as digital signal processing (DSP), data reduction, or the squeezing down of data, is applied to audio and video signals in order to save storage space and allows for a portion of the signal to be transmitted, yet the receiver can "reconstruct" the entire signal

(Cole, 1992, p. 51). Compressed signals use a minimal amount of spectrum space, conserve bandwidth, and allow use of multiple audio and video services within the available spectrum space (Hoffner, 1994). A tremendous amount of audio or video "is redundant" or "repeated, repetitive, superfluous, duplicated, exceeding what is necessary" (Weiss, 1992, p. 35). Because "the human eye/brain combination has limitations to what it can perceive . . . it is possible to greatly reduce the amount of information that must be transmitted" from a microphone or camera to a receiver to persuade someone that he/she is seeing or hearing what a microphone or camera picked up (p. 35). Although "large portions of the audio signal can be discarded as irrelevant to the ears," the drawback is that quality degradation can "be detected down the line" (Treasure, 1992, p. 5). The International Radio Consultative Committee's (CCIR) engineering task group TG-10/2 determined that digital audio can only "withstand so many compression schemes before compact-disc quality becomes sub-broadcast quality" (Compressing business, 1993, p. 5).

Digital compression of video works similarly to audio compression, as the TV set, like the radio receiver, is able to correct errors and to restore a noisy signal to its proper modulation level (Mannes, 1991, p. 30). Digital channels take up less space on satellite transponders, which allows cable and direct broadcast satellites (DBS), like digital audio radio services (DARS), to distribute more channels and more diverse programming (Compression: Changing, 1990).

DIGITAL AUDIO BROADCASTING (DAB)

DAB has had numerous acronyms. At one point, the FCC said DAB would apply to terrestrial delivery and DARS (digital audio radio services) would refer to both satellite and terrestrial DAB. SDARS would include only satellite-delivered digital audio. The Electronic Industries Association (EIA) preferred DAR, for digital audio radio, which should not be confused with DARS. DRB, meaning digital radio broadcasting, was contributed by CDRB (the Committee for Digital Radio Broadcasting). Another acronym from an unknown source was BSS, broadcast satellite service. Many observers wondered what was wrong with plain DAB (Zavistovich, 1992).

DAB can be delivered via in band, on channel (IBOC), through in band/adjacent channel (IBAC), or by satellite digital audio radio services (SDARS). IBOC can be facilitated by using existing terrestrial AM and FM spectrum, but the major benefit is that digital compression allows more stations to utilize the same amount of spectrum space normally occupied by one analog signal. Terrestrial DAB is transmitted from towers like traditional radio, but as pointed out, the main difference is in the economical use of spectrum. A powerful analog signal can use 100 thousand watts, but digital delivery can reduce that to 1 thousand watts (Pohlmann, 1992, p. 64). IBAC is very similar to IBOC, but uses blank spaces between existing AM and FM frequencies (Mitchell, 1993). SDARS is totally different, as the signal is sent via DBS directly to consumers' receivers.

The FCC (1990) opened its DAB proceeding on August 1, 1990, but the technology was already under development in Europe in the latter part of the 1980s. The European Broadcasting Union (EBU) was working on its own terrestrial DAB system known as Eureka-147 for deployment on a new band (NB). American broadcasters were not keen to abandon the AM and FM bands for NB, despite the fact that the National Association of Broadcasters (NAB) initially favored NB as an equalizer for AM and FM broadcasters. NAB favored L-band and EBU's Eureka-147. Angry broadcasters persuaded NAB to support IBOC DAB transmitted via an American-made system. A consortium called USA Digital Radio (USADR) was formed in 1991 by CBS, Westinghouse Electric Corporation, and Gannett Company to develop an AM and FM IBOC system called Project Acorn.

Project Acorn and Eureka-147 employed MUSICAM (Masking pattern Universal Sub-band Integrated Coding And Multiplexing) compression, which was expected to become the de facto compression standard. But, soon after AT&T/Bell Labs merged with Lucent Technologies to form Lucent Digital Radio (LDR), they created a new compression scheme called Perceptual Audio Coding (PAC). PAC employs psychoacoustic modeling, which is "a mathematical model of how the human auditory system hears sound" (LDR, 1998). Like other compression methods, PAC did not transmit unnecessary or redundant information.

Naturally, DAB promises consumers the "next plateau of audio performance" (Pai, 1997, p. 14), offering improved sound and greater fidelity than analog AM and FM broadcasting. Digital delivery improves stereo separation and reduces interference and signal fading more than analog delivery does (Gatski, 1990). Audio signals are very susceptible to interference during mobile reception. Most of that interference is due to multipath reception, which simply is a radio signal's susceptibility to interference from other signals bouncing around the airwaves. As most radio listeners know, buildings, hills, and other large obstacles can obstruct radio signals. Because the signal begins bouncing around, there is overlap of direct and reflected signal components that arrive at the receiver at different times. Unlike analog broadcasting, digital receivers choose the better of those signals whether they are reflected or direct.

DIGITAL TELEVISION (DTV)

As with radio, over the years there have been various attempts to improve U.S. television. DTV emerged during the pursuit of HDTV in the early 1980s, and new abbreviations and terminology appeared at lightning speed. We began to hear new terminology such as DTV, advanced television (ATV), standard definition television (SDTV), improved-definition television (IDTV), ultradefinition television (UDTV), and enhanced-definition television (EDTV). Given the appeal of the word "digital," it is no small wonder that DTV was the term that has endured.

DTV is the third major enhancement for television since 1941, when the National Television Systems Committee (NTSC) established its standard for picture resolution. Two other major milestones were color TV in 1954 and MTS in 1984. DTV and HDTV are frequently used interchangeably, but they are different. DTV includes digital delivery of video, audio, and other auxiliary services. HDTV has primarily referred to picture resolution.

From the beginning of its proceeding to improve television delivery in the late 1980s, the Federal Communications Commission (FCC) called the improvement ATV. To the FCC, ATV seemed more apropos than HDTV because it would encompass any advances in television technology. Broadcasters and consumers eagerly anticipated an analog system with 1,125 lines of resolution. The NTSC standard is 525 lines. With the entry of digital delivery, however, lines of resolution and frames per second (fps) as measures of video quality became almost moot. Talk turned to pixels, a computer industry term for picture elements representing the smallest detail that can be produced in an image. SDTV, the digital equivalent to the NTSC analog signal, contains about 300,000 pixels per frame at 30 fps, but HDTV contains at least 2 million pixels (Ranada, 1999). Robin (1998) explained: "The computer industry uses the term *vertical resolution* when referring to the number of active lines per picture and *horizontal resolution* when referring to the number of active pixels per line. Resolution can be misleading, especially with several different computer resolution formats. Among them are: 640 x 480, 800 x 600, 1,024 x 768 and 1,280 x 1,024" (p. 46). Picture definition, or resolution, began to change from the number of lines on the screen to the quantity of visible detail and to the quality of audio. A DTV picture can contain up to five times more detail and ten times more color information than found in analog signals. Audio for DTV is, of course, CD-quality, which makes it much more desirable to techno-junkies than FM-quality MTS.

Digital's entry into the TV industry merely made matters more confusing than they already were with HDTV. It is very important to understand the sentence that follows. DTV can be HDTV or SDTV, but not all DTV is HDTV. Specifically, HDTV refers to six operating modes of the DTV system supporting CD-quality sound, higher-resolution video with a widescreen 16:9 aspect ratio image, and scanning at 1,080 or 720 lines. Twelve other DTV modes produce SDTV that, with its lower-resolution scanning at 480 lines, is comparable to NTSC video. As Masters (2001) noted that resolution has become as big a buzzword as has digital:

> Resolution is one of those ultra-popular specs that few people—or even TV manufacturers—understand completely. Vertical resolution is easy since it's fixed by the number of scan lines: 480, 720, or 1,080. Horizontal resolution should mean maximum number of alternating black and white vertical lines that can be distinguished in a portion of the screen as wide as it is high. Many resolution specs are for the whole screen width, however, and are thus overstated. If the 1,920-line spec is for the whole screen, the actual resolution would be 1,080 lines. In fact, a 1,080-line HDTV signal has 1,920 pixels per line, leading to about equal resolution horizontally and vertically. (p. 30)

Broadcasters have long shown much more interest in digital SDTV than they have for HDTV. Multiple SDTV programs can be broadcast in the same amount of spectrum space used by a single HDTV program (Ranada, 1999). In other words, broadcasters have seen more money to be made in transmitting more channels at lower resolution than they have in transmitting one channel with superior resolution.

The FCC (1990) began its ATV rulemaking proceeding in 1987, but over the years it discontinued using the term ATV in favor of DTV. Similarly to DAB, a partnership called the Advanced Television Research Consortium (ATRC), aka the Grand Alliance, was formed in 1990 by North American Philips, National Broadcasting Company (NBC), Thomson Consumer Electronics, and David Sarnoff Research Center. The consortium created the Grand Alliance DTV transmission system and worked advisory to the FCC during its proceeding. The Grand Alliance recommended standards and operating parameters for DTV transmission.

Along the way, the FCC and the Grand Alliance had to deal with computer guru Bill Gates and movie mogul Steven Spielberg, who did not want a transmission standard established for DTV. Gates and Spielberg, representing a coalition called Americans for Better Digital TV, wanted the flexibility to choose between progressive and interlaced scanning display monitors (Broadcast digital TV, 1996). Gates and Spielberg wanted to block the interlaced Grand Alliance system even though it was expected to be totally progressive within five years. Progressive scanning requires much more spectrum than interlace. Broadcasters did not trust the computer industry, whose hardware and software manufacturers were seen as greedy sorts who were less interested in the TV industry than they were in selling new hardware and software every year. TV production equipment and consumer TV sets typically last for years, unlike computer software and components (Broadcast digital TV, 1996, p. 9). Dick (1996) stated bluntly that computer makers opposed a DTV standard "because then they couldn't obsolete your TV set every 18 months" (p. 4). Without a standard, however, consumers would "be faced with confusion and worry about the ability to receive ALL digital signals or the obsolescence of the equipment they must purchase" (Reber, 1997, p. 8).

Ultimately, flexible use won out, as the FCC never sanctioned any specific transmission formats for DTV. Epstein (1997) explained: "Early sets will be able to receive and display the original formats [proposed by the Grand Alliance], and it is expected that broadcasters will broadcast signals adhering to these formats. However, since they are not contained within the standard, there is nothing to stop further formats from being introduced. This leaves open the possibility that televisions may become obsolete every few years as new image formats come into play" (p. 9). Dick (1998) concluded: "It seems to me that all of this talk about different platforms will do nothing but confuse the viewers. I guess that wouldn't be so bad if the intent was to provide the best service possible, but I suspect the format selection was a lot more political than it was analytical" (p. 10).

Does digital equal success?

Despite incorporating the word "digital" in its descriptor, there has not been a rush by consumers to embrace DTV. So far, DTV's acceptance has been as begrudging as that of color TV, MTS, and AM stereo. But, as one observer noted, "It was a 'keep-up-with-your-neighbor' mentality that pushed CDs along, and that will probably be true" with DTV before all is said and done (Powers, 1991, p. 68). Still, neither DTV nor DAB ensures that a broadcast station will be more successful than it has been with analog. Although it took the early initiative with color TV programming, NBC made little headway over competing networks. Garvey (1980) stated: "While this expansion of color programming eventually built an audience with color sets, there is no evidence that NBC gained much from its considerable investment" (p. 521). NBC had an advantage because of RCA, which owned the network. RCA produced color TV sets and was led by the genius of David Sarnoff, making it able to avoid a "chicken and egg" or "circular problem" experienced in the 1980s by AM stereo. Garvey (1980) explained: "Color made economic sense if it was viewed in the broader context of programming, audience, *and* manufacturing" (p. 523). Nonetheless, competing networks were not harmed economically by waiting to implement color (p. 523). Still, a problem for DTV and DAB is that neither has David Sarnoff and RCA to push them to prosperity. DAB and DTV have critics who question their necessity. The same was true for AM stereo, because most consumers really did not see a need for it when they already had FM stereo. Most of the criticism directed toward radio and television has had more to do with the state of programming than with audio or video quality. Consumers tend to seek out favorite stations, programming, or personalities. Nonetheless, radio and television certainly benefit when technological quality is also part of that equation.

The FCC and Regulation

AM Stereo

Unfortunately, the FCC has done nothing to help us resolve the confusion for broadcasters and consumers over all these new technologies. In fact, beginning with the AM stereo proceeding from 1977 to 1982, the FCC has contributed greatly to the chaos. The FCC opened its original AM stereo *Notice of Inquiry* (NOI) in 1977, and it ultimately allowed AM stations to begin stereo broadcasting in 1982. However, the FCC chose no standard transmission system for AM stereo as it had historically done for all other broadcast technologies. Five systems were approved to broadcast AM stereo: Kahn Communications, Inc./Hazeltine Corporation (Kahn), Belar Electronics Laboratory, Inc. (Belar), Harris Corporation (Harris), Magnavox Electronics Company (Magnavox), and Motorola, Inc. (Motorola). Engineered and designed differently, each system was incompatible with the other four. To pick up each of the five signals required five different receivers. Therefore, it was necessary to pick one of the five systems to be the AM stereo transmission standard.

With the exception of the first color TV decision and DTV, compatibility has been an important consideration in FCC proceedings. For a system to be selected, it needed to accommodate existing receivers. When the FCC sets a new transmission standard, such as color TV, FM stereo, or MTS, it is important that consumers be able to receive the new signal using their same receivers. For instance, a black-and-white TV set could still get a color program; it just would not be in color. And, a monophonic FM or TV receiver could continue to receive a monophonic signal even if the broadcast was in stereo. DTV and DAB signals, because they are digital and not analog, cannot be deciphered by older sets. The solution is to buy a new set or to buy a converter that would change the signal back to analog. For several years now, TV sets have been available that can be upgraded to digital—either SDTV or HDTV. Digital broadcasting makes true compatibility virtually moot. As a result, we have seen the rise of new terminology such as convergence and interoperability. In other words, technologies are designed to allow communication between various components.

Since the 1982 AM stereo marketplace decision, the FCC no longer appears to concern itself with compatibility. The FCC only polices technical parameters, or guidelines, for transmission. After a congressional mandate, the FCC picked Motorola's AM stereo system as a standard in 1993. Other than that, the FCC has waited for the broadcast industry to settle on its own standard. Therefore, MTS and DTV systems were recommended by consensus to the FCC. We have seen a new regulatory model created, which is known in some corners as Amstereoization (Huff, 2001).

Of course, setting standards does not necessarily mean that a technology will be successful. It does mean, however, that the technology will have a chance to succeed or fail on its own. As the AM stereo marketplace exemplified, no standard means certain failure, because neither broadcasters, receiver manufacturers, nor consumers know what to do.

AM stereo was entirely different from FM stereo. The FCC had a choice of 14 systems, nearly three times the number of AM stereo systems, but it made a decision and stuck to it. Broadcasters, receiver manufacturers, and consumers acted confidently. The FCC must have known that AM stereo would not work without a standard. Consumers heard some stations touting AM stereo broadcasts, but many did not hear them because the station may have employed the Kahn system and the listener may have had a Motorola receiver. With the first color TV standard, consumers' color sets could pick up CBS's broadcasts but could not receive other networks at all. Why? Because existing black-and-white sets were incompatible with the CBS system. Ultimately, the FCC was forced into choosing RCA's system that was compatible with consumers' black-and-white TV sets. Certainly, most people would have been interested in color broadcasts. To some it was just too confusing, and to most it just was not worth the hassle. As Huff (2001) noted: "In a market in which vast numbers of consumers cannot set the clocks on their VCRs, much less program them, were these same people supposed to figure out which AM stereo system a station uses" (p. 6)?

It becomes very obvious that FCC procedure and policy in setting transmission standards is of paramount importance. FCC policy and procedure have greatly and differently affected the introduction of color TV, FM stereo, AM stereo, MTS, DAB, and DTV. Without equipment availability and saturation, consumer education and promotion are virtually impossible. AM stereo was clearly a technology that never had a prayer to succeed.

Programming

A very important question begs an answer, or at the very least, some serious consideration and thought. Are TV viewers and radio listeners dissatisfied with those media? Are consumers looking for a better picture and better sound quality? Given a choice, would consumers prefer better audio and video, or would they prefer better and more diverse programming options? Doyle (1992) responded: "If you walk down the street and ask somebody what's wrong with television, you're not going to find anybody who says: Resolution" (p. 165). As mentioned, the digital perception may give some impetus to DTV and DAB. But, just how much interest will be generated among consumers? Beacham (1997a) cautioned that it cannot be assumed consumers "will automatically pay a significant premium for a TV set that receives a slightly better over-the-air picture" (p. 27). Indeed, he believed, DTV "could be a big-time consumer electronics flop" (p. 27). For instance, home electronics consumers chose the Video Home System (VHS) format over laser discs, and the jury is still out on the VHS versus DVD competition.

It can be argued that acceptance of any new electronic media technology depends tremendously on the programming offered. It is very doubtful that the average consumer can detect much difference between the present TV signal and the new DTV one. When a technology first emerges, it typically generates a degree of interest, but that initial interest does not necessarily translate into success in the long haul. There was great interest in AM stereo in 1982, but over the years the technology survives in only a few places. Indeed, there have been only five consumer electronic media technologies that have achieved or surpassed 85 percent penetration. Those technologies are telephones, radio receivers, monochrome TV sets, color TV sets, and VCRs. Although black-and-white TV offered pictures that radio could not, it did not achieve 50 percent saturation for eight years. It took color TV 17 years to climb to 50 percent saturation. Color was nice, but the high cost and limited programming could not entice viewers to make the switch. Although TV (monochrome and color combined) achieved 98 percent penetration in 1978, color TV reached 83 percent saturation 25 years after its introduction. Garvey (1980) said TV's problem "was exacerbated by the high price of color sets, equipment, and programming" (p. 515).

DTV faces a similar reality. As a point of comparison, VCRs were on the market for 20 years before 85 percent penetration was reached in 1995. Fast? Maybe not, but VCRs hold the record for the shortest time needed to make it to 85 percent saturation. Given that same fast track, it would take DTV until 2017 to reach 85 percent penetration (Schubin, 1998). CD players have not yet made it to 85 percent, and after 50 years cable TV only climbed to 67 percent in 1998.

Upon television's introduction, it seemed much interest was sparked because it had pictures that the radio did not. Although many eagerly anticipated color TV, the level of intrigue was not as great as it had been for the first black-and-white sets. With all the rapid technological development and entertainment sources that exist now, it takes much more innovation to interest consumers who have developed a higher level of sophistication. In regard to DTV, Pensinger (1997) asked: "Once we've built it, will they come" (p. 18)? Some observers believe quality programming is a better enticement for consumers than technological innovation (Husted, 1998, p. C-1; Krantz, 1997, p. 69; Seeing the future, 1991, p. 5). Although some viewers will watch anything, Dick (1998) said for most consumers, "It's the programming, stupid!" (p. 10).

A similar scenario appears to be true for DAB. There are those in the electronic media industry who remain "unconvinced that the consumer will go out and buy special equipment to receive digital quality when an alternative is available to them in FM stereo" (Looking into, 1992, p. 50). DAB's success will depend greatly on penetration into the automotive market (NAB poised, 1996, p. 55). Sirius Satellite Radio (Sirius) and XM Satellite Radio (XM), two ambitious SDARS services, debuted in 2001. Although separate entities, Sirius and XM allied to compete with traditional AM and FM radio services for mobile listeners. Unlike DAB, Sirius and XM are subscription services offering 100 channels for about $10 per month, so only time will tell how much headway they can make against free radio. If cable and satellite TV are indicators, radio could have tough times ahead.

So, even though DAB has not been initiated in the United States, it is already threatened by SDARS delivery services. Whenever the FCC approves DAB, there will be a lengthy transition period. The automobile industry has typically embraced new radio technology, as it did AM stereo, but it could take half a decade to get DAB receivers into new cars after a standard is set. That assumes, of course, that a standard will be set, and there is no guarantee anymore with the FCC. In the meantime, the major auto companies and receiver manufacturers are beginning to install SDARS receivers for Sirius and XM. Will these same manufacturers be motivated to bother with DAB? Will consumers, who will by then be exposed to this new digital radio medium of SDARs, see any need for DAB?

As with DAB, there is little doubt that consumers will play a pivotal role in determining DTV's success or failure. Few observers are convinced DAB and DTV will experience immediate success, and there is much history to support such a position. Although there is credence in looking at adoption histories of previous broadcast technologies, the Congressional Budget Office (CBO) said caution should be taken in making comparisons between color TV and DTV. CBO (1999) cited the ability of existing black-and-white sets to receive the color broadcasts. Color programming "could be viewed, at a lesser quality but without additional expense, on existing black-and-white sets–which reduced the incentive to buy a new color one." The scenario was very similar to existing FM radios decoding stereo broadcasts and converting them to monophonic. CBO (1999) added: "Perhaps the largest difference between the introduction of digital TV and other consumer electronics devices is that digital TV is being introduced into the marketplace as a public policy, enjoying government mandates such as technology standards (absent in the introduction of VCRs) and required broadcasting (absent in the introduction of color TV)." CBO suggested the government-mandated deadline "should help speed" DTV's "adoption compared with a product whose introduction is completely driven by the market."

No one has bothered to ask consumers if they want DTV or DAB or any other technology for that matter. Indeed, TV technology has endured very well. Are DTV and DAB necessary, or even suitable, alternatives to traditional TV broadcasting? Perhaps DTV and DAB are merely reactions to an ever-changing society. Whitaker (1998) noted that "consumers have demonstrated an insatiable appetite for new electronic gadgets," so "as consumers branch out from traditional, over-the-air entertainment sources to other media, broadcasters are playing a whole new ball game" (p. 68).

GLOSSARY

ACATS (Advisory Committee on Advanced Television Service): FCC Chairman Dennis Patrick formed ACATS in 1987 to research HDTV and to recommend a system. Patrick appointed former FCC Chair Richard Wiley as head. Patrick wanted ACATS to establish an evolution-friendly HDTV standard.

ACTV (Advanced Compatible Television): An analog HDTV transmission system cosponsored by NBC (National Broadcasting Company), Thomson Consumer Electronics, and David Sarnoff Research Center. See also ATRC.

aspect ratio image: The TV screen has a ratio of 4:3. The Working Group on High-Definition Electronic Production, established by the Society of Motion Picture and Television Engineers (SMPTE), determined that a wider screen ratio for HDTV of 16:9, as used in the film industry, would waste less image space. The 16:9 ratio is exactly one-third wider than 4:3. Some broadcasters and many filmmakers were more interested in screen size than resolution. Widescreen sets became available in 1993 to consumers, who seemed more interested in the 16:9 ratio than resolution. Movie producers of the 1950s used widescreen formats as an attractive alternative to the nearly square TV picture, but it took TV set designers four decades to respond. Programs shot in 16:9 ratio could more easily adapt to 4:3 in the present than to convert 4:3 to 16:9 in the future.

ATRC (Advanced Television Research Consortium, aka the Grand Alliance): On January 25, 1990, North American Philips agreed to work with ACTV systems cosponsors NBC, Thomson Consumer Electronics, and David Sarnoff Research Center. They formed ATRC with the goal of establishing an HDTV system standard. They wanted to develop a simulcast system broadcasting two separate signals, as opposed to pursuing previous plans for an augmentation system. Simulcasting meant one signal would be sent as HDTV and the other as NTSC, allowing a transition period until such time that all signals would be HDTV. Augmentation meant HDTV and NTSC would be broadcast over two channels, but the system did not allow for total HDTV conversion. Until their agreement, North American Philips and others had disagreed on augmentation and simulcasting. North American Philips preferred augmentation and had been opposed to NBC's and Sarnoff's preference for a gradual transition to HDTV through simulcasting. North American Philips wanted immediate conversion to HDTV. The consortium was a major step toward making HDTV a reality, and it was hoped other proponents would join them in their "Grand Alliance."

ATSC (Advanced Television Systems Committee): FCC subcommittee designated to establish an ATV standard.

ATTC (Advanced Television Test Center): An HDTV testing facility established in 1990, which was capable of testing one system while setting up or breaking down another. Each HDTV system proponent was assigned office space adjacent to the laboratories.

ATV (advanced television): In the beginning, the FCC preferred the all-inclusive term ATV, meaning spectrum could be used to broadcast extra SDTV channels, HDTV, or other digital services. When digital technology was introduced, the FCC changed from ATV to DTV.

bandwidth: The range between the upper and lower limiting frequencies and the width of a band of frequencies, and is the maximum amount of digital data capable of transmission or storage.

BSS (broadcast satellite service): An acronym for SDARS from an unknown source.

BTSC (Broadcast Television Systems Committee): EIA's committee formed to study multichannel television sound, or MTS. MTS is better known as TV stereo.

CBO: Congressional Budget Office.

compression: Also known as DSP, data reduction, or the squeezing down of data, compression is applied to audio and video signals in order to save storage space. Digital transmissions could then work within limited bandwidth for both satellite and terrestrial delivery. A compressed signal uses minimal spectrum space, conserves bandwidth, and allows use of multiple audio and video services within the available spectrum space.

C-QUAM: Motorola's compatible-quadrature amplitude modulation AM stereo transmission system.

DAB (digital audio broadcasting): The FCC determined that DAB would apply only to terrestrial delivery of digital radio.

DAR (digital audio radio): EIA (Electronic Industries Association) used DAR as an acronym for digital radio. It should not be confused with DARS.

DARS (digital audio radio service): The FCC determined that DARS would refer to both satellite and terrestrial digital radio.

DBS (direct broadcast satellite): The audio and/or video signal is sent directly from the satellite orbiting the earth to the home consumer. Other satellite systems send the television signal to the "head ends" of cable television companies for distribution to the consumer.

DCR (Digital Cable Radio): An alternative to DAB, DCR is distributed though cable TV systems. DCR is not limited to subscription music channels, as some local radio stations are available.

DRB (digital radio broadcasting): DRB was an acronym for DAB contributed by CDRB (the Committee for Digital Radio Broadcasting).

DSP (digital signal processing): Compression of data in an audio or video signal.

DSR (digital satellite radio): A digital audio delivery alternative that has been available for years, but should not be confused with SDARS. DSR is not broadcast and is not mobile or portable. DSR, developed in Germany, is a subscription service received like satellite television via a dish aimed at the orbiting satellite.

DTV (digital television): DTV can be HDTV or SDTV, but not all DTV is HDTV.

DVD (digital versatile discs): It should be noted that, due to problems with trademarking, DVD no longer stands for anything because none of the acronyms was trademarked. Some people refer to it as digital videodiscs, but during development it was called digital versatile discs. DVD was modeled after CDs.

EDTV (enhanced-definition television): Originally, the FCC wanted to phase in DTV by allowing broadcasting to continue on existing transmitters and receivers. HDTV would begin on new channels. As a result, the FCC held the possibility of adopting an EDTV system that would be compatible with NTSC sets while providing at least some of the enhancements of HDTV. Proponents believed EDTV could ease the transition for consumers by providing an economical bridge to HDTV.

FCC: Federal Communications Commission.

HDTV (high-definition television): HDTV refers to six operating modes of the DTV system supporting CD-quality sound, higher-resolution video with a widescreen 16:9 aspect ratio image, and scanning at 1,080 or 720 lines.

IBAC (in-band, adjacent channel): A form of DAB that uses existing AM and FM bands. DAB is broadcast in blank spaces between those frequencies.

IBOC (in-band, on-channel): A form of DAB designed to allow existing AM and FM stations to use their spectrum to transmit analog simultaneously with digital signals. Broadcasters could convert from analog to digital radio without service disruption while maintaining current dial positions.

IDTV (improved-definition television): Another acronym for a television signal delivering better resolution.

ISM: Kahn's independent sideband modulation AM stereo transmission system.

JPEG (Joint Photographic Experts Group): A professional organization that developed a digital compression standard for still photography.

KCI (Kahn Communications, Inc.): A company that designed the proposed independent sideband modulation AM stereo transmission system.

L-band: A terrestrial radio band (like S-band) other than traditional AM and FM spectrum implemented in parts of the world for the deployment of DAB. L-band is a region of spectrum at 1,500 MHz (1.5 GHz) that is used in the United States for military testing of aeronautical devices.

LDR (Lucent Digital Radio): A company formed by the merger of AT&T/Bell Labs and Lucent Technologies. LDR entered the DAB process after USADR, but caught up quickly. LDR developed a digital audio compression scheme called PAC.

MPEG (Moving Picture Experts Group): A professional organization that developed a compression standard for digital television.

MTS (multichannel television sound aka TV stereo): The FCC (1977) defined MTS as "Any system of aural transmission that utilizes aural baseband operation between 15 kHz and 120 kHz to convey information or that encodes digital information in the video portion of the television that is intended to be decoded as audio information."

MUSICAM (Masking pattern Universal Sub-band Integrated Coding And Multiplexing): A digital audio compression scheme that is expected to become the de facto compression standard.

NAB: National Association of Broadcasters.

NB (new band): A terrestrial band, such as L-band or S-band, for the deployment of DAB other than traditional AM and FM spectrum.

NTSC (National Television System Committee): A U.S. committee put together in 1953 to set a color TV transmission standard. Early color TV transmission was inconsistent, which led to the nickname "Never Twice the Same Color" (Ranada, 1999, p. 93).

PAC (perceptual audio coding): A digital audio compression scheme developed by LDR that "uses psychoacoustic modeling—that is, a mathematical model of how the human auditory system hears sound—to filter out irrelevant information in an audio signal" (LDR, 1998).

PAL (phase alternation line): Along with SECAM, one of Europe's two incompatible terrestrial color TV transmission standards. Each has 625 lines and 50 cycles, but only France employs SECAM. PAL is employed in Western Europe, China, and most of the world other than North and Central America, Japan, and South Korea. PAL was developed in the 1960s by Germany's Telefunken to eliminate color problems experienced in long transmission links.

SB-ADPCM (Sub Band-Adaptive Pulse Code Modulation): A Swiss-designed digital audio compression scheme.

S-band: A terrestrial radio band (like L-band) other than traditional AM and FM spectrum implemented in parts of the world for the deployment of DAB. The U.S. ultimately reserved L-band for aeronautical use and assigned S-band (2,300 MHz, 2.3 GHz) for SDARS.

SCA (Subsidiary Communications Authorizations): FM stereo's basic groundwork was laid in 1955, when the FCC (1955, 1958) allowed FM stations to initiate multiplexing, the act of one transmitter sending out two simultaneous, but possibly different, signals. Originally, SCA were intended for nonbroadcast services, such as piped-in music often heard in medical facilities, traffic light control, paging services, and operations directed toward specific interests. The FCC also listed stereo as an SCA service.

scanning: *Progressive scanning*, primarily preferred for data and text displays, is used by computer monitors that transmit information to the screen in one pass. All lines are scanned in progression from top to bottom, from 1 to 525. NTSC video employs *interlaced scanning* in two passes creating 525 lines of resolution and 60 cycles. An electron gun alternately scans every other line on its first pass, or lines 1, 3, 5, 7, 9, 11, and so on. On the second pass, every line is scanned, including those already scanned. Interlaced computer pictures flicker, but TV does not. Progressive scanning requires much more spectrum than interlace (Barry, 1997, 1998).

SDARS (satellite digital audio radio service): The FCC determined that SDARS would refer only to satellite-delivered digital radio.

SDTV (standard definition television): A digital television system in which the quality is approximately equivalent to that of NTSC. Twelve other DTV modes produce SDTV that, with its lower-resolution scanning at 480 lines, is comparable to NTSC video. SDTV is a subcategory of the ATSC's HDTV proposal. Broadcasters' interest was piqued because several SDTV programs can be broadcast in spectrum space used by a single HDTV program.

SECAM (*séquential colour avec memorie,* or sequential color with memory): Along with PAL, one of Europe's two incompatible terrestrial color TV transmission standards. Each has 625 lines and 50 cycles. Only France uses SECAM.

Sirius (Sirius Satellite Radio Inc.): An SDARS distributor that began services in 2001. Until 1999, it was known as CD Radio, Inc.

UDTV (ultra-definition television): Another acronym for a television signal delivering better resolution.

USADR (USA Digital Radio): A partnership established in 1991 by CBS, Westinghouse Electric Corporation, and Gannett Company to develop IBOC DAB for AM and FM.

VCR: Videocassette recorder

VTR: Videotape recorder

XM (XM Satellite Radio Inc.): An SDARS distributor that began services in 2001. Until 1998, it was known as American Mobile Radio Corporation (AMRC).

REFERENCES

Barry, J. (1997, April). HDTV gets closer. *Stereo Review*, pp. 18, 20.
Barry, J. (1998, July). Digital TV demystified. *Stereo Review*, pp. 29-31.
Beacham, F. (1997a, March 13). DTV War: What is it good for? *TV Technology*, p. 27.
Beacham, F. (1997b, July 3). In praise of democratic design. *TV Technology*, pp. 22, 24.
Broadcast digital TV: A new star wars? (1996, August). *Advanced Imaging*, p. 9.
Cole, A. (1992, March). Compression race heating up. *Computer Video Technology*, pp. 51, 53.
Compressing business awaits. (1993, May 26). *Radio World*, p. 5.
Compression: Changing the world of television? (1990, June 11). *Broadcasting*, pp. 68, 70–71.
Congressional Budget Office. (1999, September). Completing the transition to digital television. http://www.cbo.gov/showdoc.cfm?index=1544&sequence=0&from=1
Diamond, A. & Sneegas, J. (1992, October 26). Monday memo. *Broadcasting*, p. 20.
Dick, B. (1996, August). Buy this baby, and it'll last you a lifetime. *Broadcast Engineering*, p. 4.
Dick, B. (1998, May). My digital is better than your digital. *Broadcast Engineering*, p. 10.

Doyle, M. (1992). *The future of television: A global overview of programming, advertising, technology and growth*. Lincolnwood, IL: NTC Publishing Group.

Epstein, S. (1997, January). The new HDTV standard. *Video Systems*, p. 9.

FCC. (1955, March 25). Part 3—Radio broadcast services: Miscellaneous amendments (Docket No. 10832; FCC 55-340). Washington, DC: Federal Register, pp. 1821–1826.

FCC. (1958, July 11). FM broadcast stations: Notice of inquiry regarding specified non-broadcast activities on a multiplex basis (Docket No. 12517; FCC 58-636). Washington, DC: Federal Register, pp. 5284–5285.

FCC. (1977, July 7). AM stereophonic broadcasting: Inquiry (Docket No. 21313; FCC 77-445). Washington, DC: Federal Register, pp. 34910–34913.

FCC. (1990, September 26). Advanced television systems and their impact on the existing television broadcast service: Policy Decision (Docket No. 87-268; FCC 90-295). Washington, DC: Federal Register 55 FR 39275.

Feedback. (2001, July/August). *Sound & Vision*, pp. 18, 20.

Forrest, J. (1994, March). World satellite market driven by compression. *TV Technology*, p. 5.

Garvey, D.E. (1980, Fall). Introducing color television: The audience and programming problem. *Journal of Broadcasting*, *24* (4), 515–525.

Gates, B., Myhrvold, N., & Rinearson, P. (1995). *The road ahead*. New York: Viking Penguin.

Gatski, J. (1990, May 9). DAB shows future of radio. *Radio World*, p. 15.

Haber, A. (1996, September 4). Transmitter manufacturers eye DAB progress. *Radio World*, pp. 19, 63.

Hahn, K. (1994, February). What defines 'digital' products? *TV Technology*, pp. 27, 54.

Hirsch, J. (1992, October). The quest for audio perfection. *Stereo Review*, p. 26.

Hoffner, R. (1994, February). Examining digital audio broadcast. *TV Technology*, p. 31.

Huff, W.A.K. (2001). *Regulating the Future: Broadcasting Technology and Governmental Control*. Westport, CT: Greenwood Press.

Husted, B. (1998, January 11). Fine-tuning HDTV. *The Atlanta Journal-Constitution*, p. C-1.

Krantz, M. (1997, April 14). A tube for tomorrow. *Time*, p. 69.

Looking into radio's crystal ball. (1992, September 7). *Broadcasting*, pp. 42, 47-48, 50.

Lucent Digital Radio. (1998, May 15). Lucent Technologies' Perceptual Audio Coder receives highest rating in independent test [Press Release]. http://Lucent.PACven051598.htm

Mannes, G. (1991, March). TV-by-numbers. *Video Review*, pp. 28–31.

Master, I. (2001, July/August). Q & A. *Sound & Vision*, pp. 30–31.

Mirabito, M.A.M. (1994) (2nd ed.). *The new communications technologies*. Boston: Focal Press.

Mitchell, P.W. (1993, November). Digital on the air. *Stereo Review*, pp. 103–106, 108, 110.

NAB poised to push IBOC. (1996, October 2). *Radio World*, pp. 1, 55.

Nelson, R. (1993, November). Swept away by the digital age. *Popular Science*, pp. 92–94, 96–97, 107, 110.

Pai, S. (1997, November 26). IBOC: Two takes on digital radio effort. *Radio World*, pp. 14, 16.
Pensinger, G. (1997, July). DTV broadcaster: The transition. *Television Broadcast*, p. 18.
Pohlmann, K.C. (1992, October). The sound of things to come. *Stereo Review*, pp. 62–64, 66, 68.
Powers offers HD views. (1991, March). *TV Technology*, p. 68.
Ranada, D. (1999, May). Survival guide: Deciphering the technobabble. *Stereo Review's Sound & Vision*, pp. 92–94.
Reber, G. (1997, March). Without a HDTV standard. *Widescreen Review*, pp. 8–14.
Robin, M. (1998, April). Digital resolution. *Broadcast Engineering*, pp. 44–46, 48.
Schubin, M. (1998, November). Forget November '98; Check out March '54. *Television Broadcast*, p. 47.
Seeing the future. (1991, April). *TV Technology*, p. 5.
Treasure or terror? (1992, December 9). *Radio World*, p. 5.
Weiss, S.M. (1992, October). Examining compression techniques. *TV Technology*, pp. 13–14.
Whitaker, J. (1998, March). The new definition of TV. *Video Systems*, pp. 65–66, 68.
Zavistovich, A. (1992, November 4). Alphabet soup and DAB. *Radio World*, p. 4.

CHAPTER 2

Old and New Teaching Skills in the Digital Age: A Layperson's Guide

Matt Jenkins

So your colleagues are jumping on the digital bandwagon. They are lusting after the latest in new DV equipment and are using such strange terms as "IEEE 1394," "wave file," and "compression." The last time you heard the term "compression" was when your mechanic sadly informed you the pistons in your car's engine no longer had compression and it was going to cost a lot of money to repair. And honestly, you didn't know what it meant then either.

These same colleagues are excitedly looking to you to teach students how to use this new equipment. What happened to the mantra that all you needed to teach production was a beat-up beta cam and a three-color camera TV studio?

It can be embarrassing to be the nonlinear editing instructor when your students know more about the technology than you do. Actually, it isn't all that bad, and the saying about the beta cam and three-color camera is still true. Teaching with digital equipment is similar to teaching production with analog equipment.

The purpose of this chapter is to point out what production skills can be transferred from analog to digital production. Also, the chapter contains a discussion regarding new concepts to be included in teaching with digital equipment. However, in writing this chapter, I have made the assumption that the reader has a rudimentary understanding of television production. This chapter should be considered a summary or starting point for understanding concepts and skills to be taught, and hopefully will help you to avoid some digitally-driven anxiety.

First let's examine some basic terms.

ANALOG VERSUS DIGITAL

All sorts of definitions exist for analog and digital signals. Comparisons from ramps to thermometers are used to describe these signals. Simply put, when an analog signal is recorded, it is recorded as an electrical signal, including interference (more commonly referred to as noise) that exists in our environment. Sources of electrical noise include such things as sunspots, washing machines, electrical power lines, and telephones.

A digital signal translates the electrical video or audio signal into a signal defined by a binary code. The electrical signal is examined and its various parts are assigned specific addresses made up of electronic zeros and ones.

Every time an analog signal is copied, the noise is copied and amplified, resulting in a lower-quality copy. The digital signal, when copied, simply copies the binary-coded signal and ignores the noise, allowing the copy to be virtually the same quality as the original.

COMPRESSION

The video signal is one continuous stream. It continuously flows through the coaxial cable or it flows through the electromagnetic spectrum from the satellite uplink to the satellite and down to the satellite receiver dish. The broadcast signal travels from the television station's antenna to the rabbit ear antenna on the back of the television set. However, try to send a television signal down a telephone line and there are problems.

Telephone signals and data signals are grouped in "clumps," or groups, of information that travel through the wire. Since the information does not move in a continuous flow, a lot of information can also fit in the spaces between the other clumps of information traveling in the channel.

In order to have a television signal be usable on the Internet and be able to travel in that environment, the signal must be changed from a continuous signal into one made up of clumps. This is done through compression. Compression is the removal of repeated or less important information from the signal. The higher the compression, the more removal of signal and the lessening of picture quality. Imagine a video of a person waving at the camera. In an uncompressed signal, the waving appears smooth and unbroken. In a compressed signal, the picture, the arm waving in particular, appears jerky and not smooth.

Compression is revisited in the sections on production values and editing.

PRODUCTION VALUES

As the instructor, you are faced with teaching a crushing amount of concepts and skills. Invariably, students return from internships or begin new jobs and are distressed as they face performing a task or executing a skill that they do not know. Unfortunately, they blame you for not teaching the task or skill they need. How do you select which concepts and skills to teach? Some instructors consider video production an art form, strictly teaching the creative aspects of the various forms of story telling. Others only consider production for its technical skills. These teachers go into great detail regarding scan rates, how the signal is recorded onto the videotape and carry on deep dark discussions over analog versus digital signals. And, of course, there are those professors who look at production from a business standpoint. They feel that they can teach students how to be producers and the students can hire the creative and technical people to accomplish the mission of their production.

Where do you stand regarding your teaching philosophy for video production? Consider what Robert Rodriguez (1996) says in his book *Rebel Without a Crew.*

> It is good to be self-sufficient in a business where people are used to relying on other people in order to get anything done. Because then people realize you can run off and make your own great film without them, and they let you have your way. That's why you need to be educated in all areas of production. If for no other reason, so no one can take advantage of you or give you false information. (p. 197)

Obviously, Rodriguez is concerned with filmmaking. But his statements are applicable to video production as well.

The more concepts and skills your students possess, the more employable they will be. They will also tend to rise more quickly through the ranks of industry. Video production is an art form, a science, and a business. While this chapter mainly focuses on the scientific aspects of video production, especially those skills needed for digital production, we should not ignore video production as an art form and as a business.

ASPECT RATIO

Aspect ratio is defined as the width and height of the television screen. The current aspect ratio for television is 4:3 units. This means the television screen is divided into four equally sized blocks across its width and three equally sized blocks along its height. As American broadcasting converts to high-definition television (HDTV), the aspect ratio is expected to change. The new aspect ratio will most likely be 16:9 units.

The 4:3 and 16:9 aspect ratios are part of a *standard,* or an agreement on how to construct an electronic visual image on the television screen. The 4:3 ratio is part of the 525 lines per frame, 30 frames per second standard. The 16:9 aspect ratio is part of the HDTV standard of 1,125 lines per frame, 60 fields per second.

What will utilizing a 16:9 aspect ratio mean? It means that video productions will more closely resemble film productions based on screen view-space, aspect ratio, and picture quality. Being able to shoot with the HDTV standard means that the pictures will have more detail. Directors will be able to use more wide-angle shots and perhaps more tracking shots than zooms. Directors will have to consider the extra information present on the screen. What will the director want the audience to pay attention to?

It means that television will no longer be considered the "close-up" medium. The less defined picture of the National Television System Committee (NTSC) standard necessitates close-ups, as detail and texture are simply not seen in long shots. The NTSC, established by the Federal Communications Commission in 1941, decided the original standard for creating the video image (Massey & Baran 2001). The standard was later amended for color broadcasting.

There are ramifications to consider when showing a 4:3 aspect ratio image on a 16:9 aspect ratio screen. When played on a 16:9 aspect ratio screen, the 4:3 aspect ratio image has borders on the left and right sides of the screen. Many viewers do not prefer this type of screen. So broadcasters may increase the size of the 4:3 image so it matches the 16:9 screen. The result will be that heads and feet will be cut off. The tendency will be to frame for the head and lose more feet. At best it is a compromise (A user's guide to aspect ratio conversion, 2001).

When commercial television begins broadcasting in HDTV using the 16:9 aspect ratio, using pan and scan or letterbox techniques will not be necessary when preparing films for broadcast. Modern movies already use an aspect ratio similar to the 16:9 aspect ratio. In order to view these movies on the traditional 4:3 television screen, a pan and scan or letterbox technique has to be applied to the movie. Pan and scan essentially means that in transferring the larger aspect ratio screen to the smaller aspect ratio screen, the view space would pan left or right to accommodate the action. Many directors shoot theatrical releases with the television market in mind and concentrate all the action in the middle of the screen so pan and scan is minimized. The problem with pan and scan is that the process adds movement to the shot that originally wasn't there.

When a movie is letterboxed, the entire picture is shrunk so the vertical edges match the vertical edges of the smaller aspect ratio screen and a thick black bar can be seen at the top and the bottom of the screen.

According to *A User's Guide to Aspect Ratio Conversion* (May, 2001), the 4:3 and 16:9 screens are not conducive to converting from one to the other. Effective wide shots in 16:9 do not work in 4:3. The director, shooting in 16:9 and concerned with image quality in 4:3, needs to remember that some 16:9 wide-angle shots will be too wide in 4:3 and flattering close ups in 4:3 will look uncomfortably close in 16:9. News broadcasters will be faced with how to handle different aspect ratios in story presentation. Sports directors, shooting in 16:9, will need to keep the 4:3 aspect ratio in mind when framing close-ups of such things as a basketball going into the net. The world is going to a 16:9 aspect ratio. It just may be a bumpy journey (A user's guide to aspect ratio conversion, 2001).

LIGHTING

Lighting should be one of the first skills taught to beginning production students. Proper lighting is critical to successful capturing of the image by the camera. When a subject is videotaped, it is the light reflected off the subject that we are interested in. Remember, no light, no picture.

Lighting has many purposes. Zettl (2000) provides three basic reasons for lighting:

1. To provide the television camera with adequate illumination so that it can see well
2. To help the viewer recognize what things and people look like and where they are in relation to one another and their immediate environment
3. To establish a general feeling and mood of the event (p. 126)

I add two more reasons for lighting:

1. To direct attention to the most important elements in the view space
2. To provide a sense of depth

We all should know three-point lighting, and most of us teach it. Basically, this formula for lighting uses three identical lighting instruments to light the subject. The brightness of each instrument is altered to create shadows on the subject that help to provide the illusion of depth. The video screen is flat, only providing height and width. This idea of creating an artificial sense of depth is important, since television is two-dimensional. Even as early as 1933, Rudolf Arnheim discussed the perception of depth and the idea of presenting a three-dimensional object on a two-dimensional plane. The sense of depth can be created through the use of camera angle, subject placement, use of background, and lighting. Using lighting to create depth adds a sense of realism to the shot and helps the audience to differentiate between objects and distances. Plus, the striving for the creation of a sense of depth in the view space is what separates those people who actively shoot and produce and are considered "professional" from those who go and buy equipment and claim they are "videographers." The concept of depth is still important for digital production.

Many professional lighting directors cringe at the mere mention of the term "three-point lighting." They don't like to light using "formula" lighting, believing it constricts creativity in lighting. However, three-point lighting is a great place to begin a discussion on lighting, especially in relation to depth.

With the advent of low-light-capable cameras, it is possible to shoot under poor lighting conditions and still achieve an image. However, low-light-capable cameras should not be used as an excuse for sloppy or nonexistent lighting. Lighting is one of the keys to a successful production. Whether shooting in analog or digital, poor lighting always ruins a good production. Great lighting always makes a mediocre production look better.

LIGHTING-CONTRAST

Analog television technology does not handle contrast well. Contrast is defined as the difference between the brightest and darkest parts of the view space.

In using new digital equipment, colleagues have remarked that black, when recorded digitally, has a grain or a "velvet" texture to it. If the contrast is greater, the darker areas tend to have a kind of grainy-gray color. Having even lighting with little contrast may still be a good idea when shooting in digital. However, this grain in the black area depends on the quality and type of the equipment used. Lighting, as done for film, uses shadows and combinations of light and dark areas to create mood, draw attention, and sufficiently represent the subject. These ideas should be discussed with the student. If shooting in digital HDTV, students will be able to provide for more contrast in the shot due to the high definition of the format.

Also, when shooting for Internet distribution, lighting is crucial for accurate picture reproduction. After the picture has been reduced in size and is compressed, being well lit allows the picture to retain quality. So lighting should still be one of the first skills taught in digital production.

CAMERA MOVEMENT

Beginning production students love to zoom, pan, and tilt the camera. It is as if they are saying, "Look what I can do!" An important part of the instruction process is to teach students to use these movements appropriately. Students should only zoom for a reason.

Here is an example of using zoom for a reason: The audience is viewing a close-up of a single, pretty flower. A slow zoom from the close-up to a wide angle reveals that the lone flower exists among the garbage and trash of a junkyard. The zoom is used to focus attention on the flower and then the zoom out introduces the viewer to the concept that the beautiful flower is growing amidst the clutter and chaos of the junkyard.

It takes a lot of practice to learn to zoom, pan, or tilt the camera smoothly and to retain a center of focus in the shot. This will only become harder with a 16:9 aspect ratio, as more information will be present in the view space. There are plenty of student examples where students poorly pan a large space. The view space reveals an aimless pan, meandering the room for something of interest. Again, the lack of focus on the part of the videographer is evident.

So it is clear that zooming, panning, or tilting should be done for such specific reasons as to draw attention or to introduce the viewer to new information. However, what about the use of these movements on a production placed on the Internet? Here is where compression rears its dubious head.

Lemm (2001) indicates that camera moves should be limited for productions distributed over the Internet. Students need to know that a lot of movement, including zooms, pans, and tilts, will look jerky, uneven, and unnatural when the video signal is compressed. This means that zooms, pans, and tilts should be used even more sparingly or perhaps not at all. The typical response then is, "Why not just have still images on the site?" The answer is, perhaps, for now that would be preferable. However, with the rapidly changing world of computer technology, by the time this book is in the reader's hands, the compression problem may be solved and broadcast-quality video may be possible.

Framing close-ups look better on the Internet than long shots, especially when the audience is viewing the piece in a small window. Compression plays havoc with detail.

LINEAR AND NONLINEAR EDITING

Linear editing is the act of editing from the start of a program to the end. The editing takes place in a line. The editor has no choice but to edit from the beginning of the program to the end. And, after the program is completed, if there are any editing changes to be made at the beginning, the program will probably have to be edited again from that point to the end of the program. Tape-based editing systems use a player and an editor machine to perform linear editing.

One of the areas in which video production has radically changed is in editing. Using a personal computer to store, manipulate, and edit video images has created a whole new set of concepts and skills. This ability to edit video in any order you choose as opposed to having to edit from beginning to end is commonly referred to as nonlinear editing. Being able to easily change the order of shots, undo mistakes, edit from any point to any point, and provide graphics easily are just a few of the advantages that justify using nonlinear editing over linear-based editing machines.

Roth (1999) indicates that a nonlinear system consists of four basic ingredients:

1. A general purpose computer
2. A video capture/output device (or IEEE 1394 port)
3. A fast hard disk drive
4. Software to capture and edit the video (p. 228)

The students' approach to editing on a computer should start with a few key ideas.

One such concept is the logging of footage. As the result of a lack of logging footage, I often find students zooming back and forth through their tapes on the tape-based editors searching for a particular video clip. There are some other problems with not logging footage. Students should know that the constant fast forwarding and rewinding while in play quickly wears out tape heads, stretches the videotape, and wastes students' editing time.

When the students are faced with inputting video or digitizing footage into a hard drive, they should understand that all the footage may not fit on the hard drive. The hard drive can fill up quickly. Hence, it is important to log the footage and only input the footage necessary for the production. Of course, the idea of logging footage is met by groans from students. But if logging, in combination with deciding the majority of editing choices, is done prior to using the editor, hard drive space will be conserved and the amount of time actually spent on the machine will be reduced.

Spending less time on the nonlinear editor may not be appealing to the students, especially when they are using the school's equipment. But when they go out to work in the professional world and editing time is sold by the hour and is expensive, the amount of time spent on the editor becomes a major factor.

Students should be taught to recognize common aspects of nonlinear editing that are transferable across platforms. Examples of common aspects include recording video into the editor, clip storage, and a time line. AVID refers to recording video as digitizing or recording, and Adobe Premiere (1995) calls it capturing. Adobe Premiere (1995) defines a video clip as

> the source material for a movie. A clip can be a movie, a still image, or an audio file. In Adobe Premiere, a clip is a pointer to a file stored on a hard disk. (p. 9)

The term "time line" seems to be generic across platforms. It generally is the section of the screen where the clips can be placed and arranged in order to complete the production.

Teaching your students to recognize common aspects will keep them from being limited to a specific platform. Depending on the type of nonlinear editor you use, you may need to select a compression rate. While selecting a higher compression rate allows you to store more information on the hard drive, it reduces picture quality, audio quality, and perhaps even screen size.

It is a joy to be able to pick any video clip that has been placed in the computer, trim the beginning and end of the clip, and simply drop it on the time line. Although it is not always recommended, the editor can begin at the end of the production and edit backward to the beginning if desired. Editing can begin before all the shooting is completed. As footage arrives to the editor, the clips can be added to the production. If the director or producer expresses displeasure at how the piece is edited, the edits can easily be changed without having to edit the entire program again, as would have to occur with linear editing.

There is an old saying about shooting with analog equipment: "Shoot for third generation." This means that the camera operator has to be sure that all precautions should be taken so the image looks as good in the third copy as it does in the first copy. With analog equipment, this is impossible to achieve. When the footage is edited and transferred to a copy to watch, it has been copied at least two times. This means that the noise has been copied at least two times and picture quality has degraded. The original raw footage is considered to be the first generation. The edited master is the second generation, and the copy that is seen is the third generation.

Using digital equipment for the most part eliminates this problem. While Roth (1999) says that it is not possible to have unlimited copies of digital signals without some degradation, students shouldn't have any problem with image quality if the digital signal is maintained until the final copy is produced. Of course, this is only true provided such concepts as quality lighting and quality audio are observed.

TO LINEAR EDIT OR NOT TO LINEAR EDIT

A debate rages among my colleagues at other universities regarding the teaching of editing. It appears that as the price of nonlinear editing equipment becomes more reasonable, these colleagues are shedding their tape-based straight cut editors as quickly as possible. Their students are only being taught editing in the nonlinear environment. I say that tape-based straight cut editing should still be taught! That is where the disagreement lies.

Picture this. You are standing in the busy newsroom of a network-owned and -operated television station in the Top Ten market of Dallas. Along one wall is a row of eight editing bays. Six of these bays contain digital tape-based straight cut editors. Two contain nonlinear systems. If you are only teaching nonlinear editing, are your graduates going to be able to easily function in this environment?

Perhaps the argument lies deeper than straight cut versus nonlinear editing. Maybe a discussion should be held on why people need a college education. What is a college education for? Is its purpose to enlighten the learner or is it to train the student for some kind of employment or career? Perhaps it is a bit of both. I know my students want jobs and want to be able to perform those jobs capably.

Also, with linear editing you can begin editing immediately. This is especially important in the time-sensitive area of news where deadlines must be met. With nonlinear editing, you must load the footage into the computer prior to commencing editing.

CABLES

Quickly, how many cables are needed to hook up a professional-grade analog VCR to a monitor? Most likely two cables, one for audio and one for video. And if you want to edit, then you must hook up more cables for the editor controller to be able to transfer the audio and video signals from the source or play deck to the record VCR. With great excitement I began to hook up our new AVID digital editor. From the digital video deck to the computer, how many cables did I need? ONE! That's right. IEEE 1394 is a standard for digital that allows the transfer of audio and video for both inputting into the computer and recording back to digital videotape as well as VCR control to all pass through one cable. Older systems required multiple cables, and cable incompatibility could be an issue. *The Videomaker Handbook* defines IEEE 1394 as

> a high-speed bus that was developed by Apple computer. It is used, among other things, to connect digital camcorders to computers. (p. 339)

Due to digital capabilities, edited video productions are now capable of being stored on CD, DVD, the Internet, digital videotape, as well as conventional analog videotape.

INTERNET DISTRIBUTION

In one sense, the Internet is the great equalizer when it comes to video distribution. Just about anyone can put up a Web site and archive or stream video for others to watch. Interested students are going to put video on the Internet. Should departments that teach "traditional" broadcasting expand into the Internet realm? Computer-based communication seems to be the technology of interest on college campuses. Perhaps Internet distribution is an area you should consider including in discussions covering production.

Technology for video on the Internet seems to change almost daily. However, according to McCleskey (2001), Internet video technology still needs more development. There has been dissatisfaction with the small screen size and low frame rates and the video quality is poor. And that is a drawback for people who want to produce broadcast-quality images.

Lemm (2001) indicates that there are two basic types of Internet video: (1) a download-and-watch video and (2) streaming video. The download-and-watch video consists of a file the Internet user copies or completely downloads and watches the video at any time. The file can be stored in a variety of formats, including MPEG, AVI, or Standard QuickTime Video. Basically, these formats are standards for storing and viewing video on the Internet. Internet users can download media players for each type of file.

Streaming video lets the Internet user watch the video as it is played from the host computer. While there is nothing to download, the picture quality is poor because of the higher compression rate (Lemm 2001).

DIGITAL SWITCHERS

Digitally based switchers, both mechanically designed and computer driven, are capable of generating digital video effects. This is a fancy term for a transition from one image to another image. Digital switchers can store all sorts of special digital video effects. Examples include page turn, ripples, mosaics, and other special pattern wipes. Some computer-based switchers allow the users to create their own transitions. Another feature is the ease with which virtually any kind of visual image, whether it be video, a freeze frame, or graphic imported from the Internet, can be combined with another visual image.

AUDIO

Often considered to be the stepchild in video production, audio has come a long way. In thumbing through an old college television production textbook that has a copyright date of 1979, I noticed that much of what was important then in audio production is still considered important today. If audio, like lighting, is weak, then the entire program suffers.

In the predigital era, having a left and right channel or a channel one and a channel two on your tape-based editor was standard. The editor could use one channel for the voice and the second channel for music or natural sound. Computer-based editing has opened up a whole new realm.

Such concepts as the definition of sound, how it travels, types of microphones and their pick-up patterns, proper microphone placement, cables, and measuring sound remain essentially the same today as yesterday. The big change comes in audio recording, storage, editing, and playback.

The old production book discusses phonograph records, reel-to-reel audiotape, multitrack recorders, audio cart machines, and audiocassettes. Except for audiocassettes, how much of this equipment is still being used? My audio production students are turning in projects recorded to zip disk, CD, and minidisc. Now audio can be recorded on a computer chip or small disk and can be opened immediately into an editing program. This allows for nonlinear audio editing in the field on a laptop computer.

When visiting a National Public Radio-affiliated radio station, I noticed only one audiotape reel-to-reel recorder. I was told that this machine was kept in order to play the occasional audiotape that came in. The station used the latest in digital technology for its productions. While many commercials are still exchanged locally on open reel tape, many nationally distributed spots are delivered by satellite.

Digital technology has radically changed the world of audio recording, storage, editing, and playback. Sound now can be recorded onto computer disk, CD, DAT (also known as digital audiotape), detachable hard drive, as well as conventional audiocassette or open reel. One advantage of upgrading to a digital audio format is the elimination of the audible hiss that is common when using analog tape-based recorders.

Audio, when recorded or captured in a computer editor, can be stored, moved, or manipulated as a file. When audio is captured with video, it is stored in the same file as the video, not a separate file. The following discussion pertains to capturing audio alone.

According to Adobe Premiere (1995), a common audio file is called AIF audio (.aif), which stands for Audio Interchange File. This is a standard type of audio file. Also, a commonly used file format is the .wav file or Windows Waveform. The .wav files are created in Microsoft Recorder for Windows or other sound editing software. Audio files, as with other kinds of computer files, are identified by the file name's extension, for example, voiceover.wav.

The editing program you use determines the kind of files you may input. While generally most audio editing programs import the commonly used file types, it is wise to check that the software is capable of importing the types of files you desire to use prior to purchase.

After the audio is captured, it can be "dragged" or moved to the editor's time line. Once placed on one or more audio tracks, the actual sine wave may be displayed. Also, volume adjustment becomes more than a mere knob. Typically referred to as "rubber band" editing, a line stretches across the track. By moving the line up or down, the volume of the track can be increased or decreased. If there are two lines in the track, one is usually for volume and the other is for cross fading between channel one and channel two or left and right channels.

While visiting a former student working as a producer for Radio Disney, I asked her to play some of her work. She deftly connected a detachable hard drive to a computer and opened up on her screen a typical one-minute promotional piece. There were at least 17 separate audio tracks displayed on the monitor. Each track held a voice, music, or a sound effect. Looking closely, I noticed there was only one piece of audio per track. So the luxury in audio editing is having the ability to use an almost unlimited number of tracks for editing purposes.

SOME FINAL THOUGHTS

Teaching production skills in the digital realm will be no harder than teaching with analog equipment. The students will need to be made aware of the changes as well as the advantages of working with digital equipment. While *Graphic Design in Television* is a book primarily concerned with television graphic construction, author Douglas Merritt makes a statement applicable to most areas of production. He says: "Creativity, story telling and imagination should always predominate" (p. 37).

This philosophy is the key to building a better graduate, a graduate who will be adaptable to future technology that has not even been thought of yet.

Take a moment and think about the equipment you used in college or as a professional. Did you use $^{3}/_{4}$-inch videotape or shoot news stories in 16 mm? While both are viable technologies, they are not comparable to today's technologies. And try now to get repair parts for the equipment! In order to remain competitive, you had to learn the new technology as it became available. You must inspire your students to be "expert" learners. Make sure they understand that in the rapid world of changing technology, the learning never ends.

As the professor, instructor, facilitator, group leader, or whatever you may be called, it is imperative to strike a balance between teaching theoretical skills and their practical application. As an example, a discussion of a camera pan would be a theoretical skill, and having the student produce a project utilizing a pan is the practical application.

Teaching students the skills for nonlinear editing and having them realize that certain aspects of nonlinear editing are common across platforms will enable your students to easily adapt to any editing platform they may encounter. Examples of common aspects include recording video into the editor, clip storage, and a time line. And a pan is a pan, whether it is done with a Panasonic S-VHS camcorder or with a Sony DVCAM.

While you can teach production using a beat-up beta cam and a three-color camera studio, remember the excitement your students will have using cutting-edge equipment. State-of-the-art equipment also helps with recruiting new students. Current students will be able to use their existing computer skills in their television projects and, yes, you will learn from them. Take the education process in small steps, and remember, do your research.

NOTE

For Dr. Myra Macon, an inspirational leader.

1. The author would like to thank the following people for their assistance in the preparation of this chapter: Stephen Adams, Assistant Professor, Communication, Cameron University, Robert Jenkins, Don Jones, NASA, and James Powell, Broadcast Engineer.

REFERENCES

A user's guide to aspect ratio conversion (May, 2001). www.snellwilcox.com/internet/reference/pdfs/arc/pdf

Adobe Premiere (1995). *Adobe premiere: Classroom in a book.* Indianapolis: Hayden Books.

Arnheim, R. (1933/1957). *Film as art.* Berkeley and Los Angeles: University of California Press.

Lemm, L. (2001). Internet video: From camcorder to computer to the world. In M. York & S. Muratore (Eds.), *The computer videomaker handbook,* 2nd ed. (pp. 303–306). Boston: Focal Press.

Massey, K. & Baran, S. (2001). *Introduction to telecommunications: Converging technologies.* Mountain View, CA: Mayfield Publishing.

McCleskey, J. (2001). Top tools for streaming internet video. In M. York & S. Muratore (Eds.), *The computer videomaker handbook,* 2nd ed. (pp. 303–306). Boston: Focal Press.

Merritt, D. (1993). *Graphic design in television.* Oxford: Focal Press.

Rodriguez, R. (1996). *Rebel without a crew or, How a 23-year-old filmmaker with $7,000 became a Hollywood player.* New York: Penguin Group.

Roth, C. (1999). *The low budget video bible: The film school edition.* 3rd ed. New York: Desktop Video Systems.

Zettl, H. (2000). *Television production handbook,* 7th ed. Belmont, CA: Wadsworth/Thomson Learning.

CHAPTER 3

Using Student Fees for Equipment Maintenance and Replacement: The University of North Texas Experience

Steve Craig

One of the most difficult challenges media educators face is obtaining adequate and reliable funding sources for the purchase and maintenance of equipment. As departments struggle to move into the digital age, this problem has become even more acute, with many universities finding that they need to replace large amounts of equipment in a relatively short time. Additionally, once purchased, digital equipment tends to become obsolete quickly. What is needed is not simply a large infusion of one-time-only cash, but an ongoing source of income that can be used for routine upkeep, upgrades, and eventual replacement. However, the existing budget systems at many academic institutions make such fiscal planning difficult, if not impossible.

In this chapter, I offer some personal observations regarding problems academic units often have with equipment budgeting, along with some suggestions for addressing these difficulties. These ideas are based on my 25 years of experience at public universities in three states, where my teaching load frequently included production courses. At two schools, I served as department chair and had firsthand experience with the negative impact on teaching caused by poor planning and inadequate equipment budgets.

Additionally, I outline how the expanded use of student fees may provide at least a partial solution to some of these funding problems. The University of North Texas (UNT) and its Department of Radio, Television and Film (RTYF) implemented such a fee system several years ago, and our experience has been very positive; in fact, I am convinced that it has made a major difference in the quality of our program.[1] While not replacing a regular budget allocation, fees can help to fill in many of the gaps. In addition, the process of collecting and using student fees imposes careful planning, budgeting, and accounting, all of which result in a more efficient operation.

THE CHRONIC PROBLEM OF EQUIPMENT FUNDING

In the radio and television industries, the need for a realistic budget to support ongoing equipment requirements is generally taken for granted, yet many colleges and universities seem ill prepared to address these same issues. Often, academic administrators are unfamiliar with the needs of our field and fail to recognize the importance of hardware in teaching production courses. Since universities spend a large percentage of their annual operating budget on salaries, a relatively small percentage is left over for equipment to be used in teaching. While the sciences can rely on research grants to help supply needed equipment, our discipline must generally rely on routine budget allocations. Unsympathetic administrators sometimes see our large equipment budget requests as frivolous or as competition for scarce dollars needed in "more traditional" disciplines.

At the same time, we who teach production courses must share at least part of the blame. Poor planning in some departments has resulted in the purchase of inappropriate, underused, or quickly superseded equipment. At some schools, intradiscipline squabbling has led administrators to conclude that equipment dollars would be better spent elsewhere. In other cases, faculty demands for new funding is so far outside the realm of possibility that administrators find it difficult to take the requests seriously.

During my career, I have found that equipment funding problems generally fall into four major categories. The use of an expanded system of student fees can help to address each of these. After examining these issues and discussing some of the more traditional ways of addressing them, I turn to a description of the fee system implemented at UNT and the benefits we have experienced through increased funding and better management.

The Overall Level of Funding for Equipment Needed to Teach Our Courses Is Inadequate

This problem sometimes stems from the fact that an institution is undergoing hard times and the entire organization's budget is inadequate. However, it also may be that upper administrators are either unaware of the level of need or are unconvinced of the unit's merit in the competition for scarce resources.

Departments can address this problem by making a concerted effort to educate administrative decision makers about the merits of the program and the level of funding required to support it. One of the often unexploited advantages of media-oriented departments is the glamorous image they derive from public enchantment with the industry. Even so, a department public relations effort may be required to keep the unit's activities in the eye and mind of administrators. Public showings of student and faculty projects, well-publicized cooperation with off-campus business and government units, and frequent contacts with alumni can all help raise the administration's awareness of a department's contributions. Faculty and student awards for productions should be touted as examples of what can be accomplished, given adequate resources.

But publicity is not enough. The administration must also be convinced that resources given to the department will be used effectively. This means departments must plan carefully and put forward a united voice concerning specific equipment needs and how those needs will address specific curriculum issues. However, departments must keep requests at levels that can be realistically expected to be met. Presenting a dean with a request for a $2 million studio facility at a time when he or she is struggling to make half that amount fund an entire college is not only bound to meet with rejection but will also color the way that any future requests are perceived. A better approach is to formulate such requests only after getting a good idea about the level of funding other units are receiving. If a sister department has recently received $200,000 to purchase an electron microscope, that amount is obviously within the realm of possibility in a request for the following year.

It is vital that the department avoid vague requests. Rather, it should present a formal plan detailing exactly why the funds are needed, how they will be spent, and the benefits expected from the new acquisitions. For large amounts, it may be politic to present a long-term plan in which a certain level of funding is requested for each of the next several years rather than in one lump sum. Although administrators are usually reluctant to make firm long-term commitments, such a plan demonstrates both the extent of the department's needs and the care with which it is planning for the future. Of course, given the nature of digital equipment, long-term plans must be flexible enough to accommodate the changes in technology and industry practice that will undoubtedly arise. Once the funds are committed, few administrators will hold a department to an exact equipment list when it can be shown that new technology serves the need better.

In making the case for new equipment, it may be advantageous for departments to document what is being done within the industry or at rival schools. For example, the fact that the Federal Communications Commission has mandated that all U.S. television stations convert to DTV can provide persuasive evidence for the need to purchase new digital equipment.

In addition to seeking out new dollars, faculty members also need to be innovative in devising strategies that ensure the most efficient use of funds and equipment already allocated. Relatively minor curriculum changes or new ways of scheduling existing facilities can often reap major benefits. For example, many schools purchase audio or video editing equipment for students to use in producing course projects. Frequently, such equipment sits idle or underused during much of the semester.

Then, a big rush hits as project due dates approach. Demand increases and students stand in line waiting for available equipment. Rather than purchase more editing stations to handle the peak demand, a revision of the course calendar can result in staggered project schedules. Another simple method of evening out demand is to make the equipment available more hours per week. Something as simple as schedule coordination between faculty members using the same equipment can result in a preservation of money to be used for other purposes.

At UNT, we recently added a new introductory course to the curriculum. Designed as a first production experience for prospective majors, it includes hands-on projects with the audio board, the television studio, a video field camera, and video and film editing suites. In order to make maximum use of limited equipment, the course is designed to be taught in three stand-alone modules, one for audio, one for studio TV, and one for single-camera production. Each module lasts four weeks and is taught by a single faculty member.[2] By offering three sections of the course with about 15 students in each section, we rotate all 45 students through all three modules during the semester. That is, the first group of students progress through the course by taking the audio module, then the studio television module, and finally the single-camera module. The second group begins with the studio television module, then goes to the single-camera module, and so on. In this way, each instructor teaches his or her module three times during the semester, and all 45 students get hands-on experience with a variety of production projects. Such scheduling makes maximum use of the equipment by spreading usage throughout the term, thus smoothing out demand and reducing the overall amount of equipment required to support the course.

Equipment Funding Levels Are Sporadic—The "Feast or Famine" Scenario

Another major problem that arises is "feast or famine" funding, the result of the difficulties some schools have in making long-range fiscal plans. At state universities, administrators must often wait for the outcome of an unpredictable legislative session before they can make spending decisions, and even then, the plans may only be viable until the start of the next state budget cycle. In such a situation, a department may experience a flood of money one year but receive little or none the next. A corollary to this practice is the last-minute spending spree, where departments are given extremely short deadlines to spend large amounts of money suddenly made available by the vagaries of the process.

Addressing the problem of sporadic funding requires continuous planning at the department level. If roller coaster budgeting cannot be avoided, faculty must anticipate by maintaining a wish list of needed items that can be periodically updated and reprioritized. The list should be as specific as possible and should include rough cost estimates and the names of potential vendors.

Inadequate Funding for Ongoing Equipment Maintenance

Most well-run commercial media production facilities have a clear, well-funded plan of regular equipment maintenance and repair. They know that routine service is essential to the reliable operation of equipment and that a good maintenance program saves money in the long run by ensuring continuous equipment availability and longer equipment life.

Yet many academic organizations fail miserably on this account. In many places, arcane accounting rules limit how units spend their budgets, so that it is possible for a department to be flush with money to purchase new equipment but destitute when it comes to funds needed to perform maintenance or upgrades. Other budget rules may require that all unspent funds be returned at the end of the fiscal year, effectively discouraging a departmental maintenance nest egg. Sometimes, poor planning by the department itself is the culprit, with pragmatic, but short-sighted, faculty members preferring to spend scarce dollars on new equipment needed today rather than set aside funds for repairs and replacements that will be required in some distant future.

Yet the advent of digital equipment has added a new urgency to the need for a good maintenance program. For example, nonlinear editing systems require not only routine maintenance of the hardware but also the software. More than ever, it is essential that departments acquiring new digital equipment routinely set aside a portion of their budget to fund service and repairs.

Failing to Fund and Plan for Ongoing Equipment Replacement

A related issue is a failure to plan for the eventual replacement of worn-out or obsolete equipment. Again, accounting regulations frequently prohibit or discourage amassing such funds. Many departments simply have no long-range plan for equipment replacement, but must wait until an item is worn out and then try to find funds from the administration to buy a new one. Although the conversion to digital equipment has given us the means in some cases to do more with less expensive equipment (as in nonlinear video editing), it also means that equipment is becoming obsolete at a much faster rate. Digital technology is essentially computer technology, and today, few computers last more than three years. It is becoming increasingly important for departments to have some sort of depreciation plan that anticipates and budgets for regular equipment replacement.[3]

At least some of these difficulties can be addressed through the use of a comprehensive system of student fees such as that used at the UNT. Not only does our fee system increase our level of funding, but it also overcomes budget restrictions associated with other funds and provides an ongoing source of money to be used for the maintenance, upgrade, and eventual replacement of equipment.

THE STUDENT FEE SOLUTION

The use of student fees to provide schools with supplementary income is not a new or unique idea. Most universities assess fees to fund many specific student services such as medical care, recreation facilities, student media, and student government. Some also use fees to pay for specific academic services such as general-access computer labs and expendable supplies used in courses.

A search for published research on the use of fees in our discipline turned up only a single, dated study (Larson, 1984), which includes only a brief discussion of fees specifically collected to support radio and television production laboratories. Reporting survey results from 220 programs drawn from a Broadcast Education Association list, Larson found that fewer than 25% of the programs responding charged any lab fees at all, and the mean for those that did was $21 for radio courses and $30 for television courses (p. 2). The study did not ascertain what types of costs the fees defrayed.

These findings support my own observation that at many schools, the rules surrounding the use of fees are so restrictive as to discourage their use. Many times the amount that may be charged is strictly limited to covering expendable supplies that students can easily purchase outside of class. Fee accounting may be so cumbersome in such circumstances that it is easier to simply require students to provide their own supplies.

However, UNT has developed a far more comprehensive system of fees. At UNT, student fees may be assessed to cover almost all the costs of equipment, maintenance, and supplies directly related to teaching a particular course. For the RTVF Department, these fees provide an important supplement to the regular operating budget, with fee-supported funds being used to repair, update, and even replace equipment used in teaching production courses.

THE UNIVERSITY OF NORTH TEXAS EXPERIENCE

The RTVF Department at UNT has an excellent reputation among area professionals for turning out well-prepared graduates. While the bachelor of arts degree program is essentially liberal arts in philosophy and curriculum, it also has a strong tradition of teaching hands-on production courses. This reputation and the university's location in the fast-growing

Dallas-Fort Worth area have resulted in strong student demand for the major. In fact, the program has grown so rapidly in recent years that enrollment caps have been implemented. The department also has advisory oversight responsibility for the 100,000-watt campus FM station and the student-operated cable TV channel. However, these units have their own budgets and equipment and are not directly integrated into the RTVF curriculum.

The department offers courses in audio, video, and 16 mm film production, as well as an extensive complement of nonproduction courses. A newly renovated building houses the department's two television studios as well as film screening rooms and audio, video, and film editing labs. A considerable amount of field production equipment is available for checkout by students enrolled in particular courses.

With the move toward more digital equipment, the department has introduced a tiered approach to the production curriculum, which provides increasingly sophisticated equipment to higher-level courses. For example, in the area of video editing, introductory students learn the basics on less expensive consumer-type nonlinear editing equipment, intermediate students use faster, more complex systems, and advanced students have access to industry-style Avid and Media 100 editors.

The UNT prides itself on its relatively low tuition—an important factor to the many UNT students who are financing some or all of their own education. In fact, the university has been included on a national list of 100 Best College Buys for four consecutive years (*Student Guide,* 1998–2001).

The goal of keeping tuition low may have had a good deal to do with the university's willingness to implement and operate our extensive system of fees. Several years ago, it was recognized that a relatively small number of the university's courses required the use of specialized equipment and materials. The fee system provides a way to support such equipment directly out of the pockets of the students who are most directly benefiting from its use.[4]

Any university course that uses specialized equipment or supplies may request to charge a "special service fee."[5] If the fee is approved, students registering for the course are billed the fee amount as part of their overall charges. All tuition and fees are collected by the university bursar and course fees are then distributed to clearly identified fee accounts managed by the appropriate department.

The amount of the course fee collected varies from course to course depending on the type and amount of equipment and materials used. In RTVF during academic year 2000–2001, these fees ranged from $20 per student in a beginning production class to $300 per student in a course in advanced film production.

Departments must file a written course fee request once per year for each course for which a fee is proposed. The request must include a detailed justification for the fee along with a detailed budget justifying the amount requested. The request is reviewed by university budget officials and may be reduced or eliminated if the justification is deemed insufficient.

Fee collection and use is governed by strict university rules enforced by periodic review by internal auditors. The basic rule is that course fees may be collected only to support the equipment, materials, and services a student will use while enrolled in a particular course. However, as detailed below, careful bookkeeping allows the department to effectively spread equipment costs over several courses and many semesters.

Student fees can be collected and used by a department to fund three major types of expenditures: *consumable supplies, salaries and wages* for personnel who directly support the course and its equipment, and for *maintenance and replacement* of the equipment.[6]

First, fees may be spent on any *consumable supplies* used in a particular course. This includes blank tape, computer diskettes, and so on. A consumable must be of a type that is normally used up by the student by the end of the semester. Perhaps most important for our department is the ability to use these funds to support our 16 mm film production courses. Fees are used to purchase raw film stock and pay for film processing and other lab services. The availability of a reliable source of funding for these items allows planning for larger, more ambitious film projects.

Fees are also used to pay the *salaries and wages* for special personnel needed to support the course. This includes the student workers who staff our equipment checkout room, as well as a full-time media services supervisor dedicated to overseeing the operation of all student labs and equipment. This employee hires and supervises student staff, does routine maintenance and small repairs, installs software, conducts inventories, installs new equipment, and makes recommendations on the repair and replacement of student equipment.[7]

Since the services provided by fee-funded staff benefit students in several different courses, a pro rata portion of each course's fees is allocated to their support. However, UNT rules specifically prohibit the expenditure of student fees to augment faculty salaries, fund research, or pay for general department administration, so all such expenses must be funded from the traditional operating budget. Similarly, any percentage of a fee-funded employee's time spent on such duties must be accounted for and funded out of regular operating accounts.

EQUIPMENT PURCHASE AND SUPPORT

Finally, student fees may be used to maintain, repair, and replace the equipment used to support courses. This includes the purchase of spare parts, maintenance contracts, software upgrades, and equipment replacement. Specifically excluded is the purchase of "new" equipment;

however, this restriction is not as severe as it first appears, since the term "new" as it is used here means new equipment that is not a direct replacement for an existing piece of old equipment. Fees may be used to purchase new items as long as it can be shown that the new equipment serves the same basic purpose as that which it replaces.

For example, as the department began its transition to digital, we were able to replace our old reel-to-reel audiotape recorders with new, dedicated computers containing audio recording and editing software. Since both old and new items were used to teach audio recording and editing, the computers were not considered "new" equipment; however, to meet the test of a replacement, the computers had to be dedicated solely to student audio work.

Additionally, fee funds may only be used to support equipment that is exclusively used by the students who pay the fees. This prohibits the loaning of equipment to anyone not enrolled in a fee-paying course, and this includes members of the faculty and staff.[8] For this reason, the department has purchased a certain amount of separate production equipment that is strictly earmarked for faculty use in research and creative activities. This equipment was purchased and is maintained out of a separate source of nonfee funds.

ACCOUNTING FOR STUDENT FEES

Properly accounting for student fees used to support equipment is probably the most complex aspect of the system. While fee rules require that funds collected in a particular course be used only to support that course, much of the RTVF equipment is shared among students enrolled in any of several different courses. A camcorder might be used by a student taking an introductory production course one day and by someone else taking an intermediate course the next. This means that detailed accounting is necessary by which the department can document how the cost of maintaining and replacing a piece of equipment is fairly divided over all the students and courses that use it. To achieve this, the department worked with UNT accounting personnel to develop a set of custom-designed procedures that ensure maximum accessibility of equipment while providing necessary documentation for audits.

It is clearly impractical to keep track of how each piece of equipment is used by each course on a day-to-day basis. Instead, the department has assigned all equipment into one of three categories: audio, video, or film, depending on how it was primarily used and corresponding to areas of the production curriculum that are the responsibility of specific faculty members. A separate fee account was then established for each area.[9]

Courses that use mainly audio equipment collect fees that are placed in the audio account and used exclusively to support the maintenance and replacement of audio items. Recently, the department developed a new introductory course designed to expose prospective majors to production practices in all three areas. Students taking the course use equipment from all three areas in labs and projects. Therefore, an appropriate proportion of the lab fees collected for this course go to each of the three equipment accounts.

One major benefit of the fee accounts is that they can provide for the eventual replacement of items of equipment that have become worn out or obsolete. Again, this requires careful documentation as to the value and age of the item. When a piece of equipment is first assigned to student use (it may either be new or used), its value is recorded and it is placed on a depreciation list. Each year the item is used, the department can write off a portion of its value. For example, if a particular computer-based video editing system costs $2,000 new and has a useful life of three years, the department can write off one-third of this amount each year, thus justifying the collection of that amount each year in fees from courses using that equipment. Noncomputer equipment has a longer depreciation life, with the exact number of years being set by administration accounting officials. This means that even though regular university funds must be used to purchase the item initially, its replacement can be funded in perpetuity by student fees.

In theory, at the end of the useful life of a piece of equipment, enough fee money will have been generated to replace the item at its original cost. In practice, things often don't quite work out this way. For one thing, the department makes an effort to keep student fees as low as possible and sometimes chooses not to request as much in fees as could be justified by the equipment depreciation figures. On the other hand, with the digital revolution, many items can now be purchased for less than the depreciated value of the equivalent analog devices they replace. Once a piece of equipment has been completely depreciated, it can be replaced; however, even if "replaced," there is no requirement that the old item be immediately discarded. We've found that solidly built nonelectronic items such as lighting instruments and film cameras can be used years beyond their ledger-page lifetime.

THE ROLE OF THE PRODUCTION COMMITTEE

An important administrative element in RTVF's effective use of student fees is the Production Committee. This group, made up of those faculty and staff members directly involved in teaching and supporting production courses, is charged with providing oversight for all production labs and equipment, formulating policies for scheduling and use, and making recommendations regarding equipment maintenance and purchases.

The Production Committee was established as a part of the department's normal faculty governance structure and is not a specific requirement of university fee rules; however, the important discussions and compromises regarding equipment that take place in regular committee meetings help ensure that fee dollars are being spent as efficiently as possible.[10] The committee itself has no direct authority to spend fee dollars—that authority rests solely with the department chair—but the committee's recommendations carry a great deal of weight.

ADDITIONAL BENEFITS OF THE FEE SYSTEM

At UNT, the fee system has clearly made a major difference in our department's ability to purchase and maintain equipment at the level required for an active production curriculum; however, we also found that the fee system provides additional benefits that might not be as readily apparent.

First, the collection and accounting for student fees has forced the department into an ongoing grass-roots budgeting process that has encouraged sound planning and fiscal responsibility. The fact that fee dollars come directly from our own students and are completely under the department's control instills in personnel a strong sense of responsibility and a real desire to squeeze every ounce of value from each dollar. The Production Committee provides an essential forum for discussion and debate about how fee dollars can be used most effectively. At the same time, because management of these dollars is the sole responsibility of the department, faculty and staff seem much more willing to address scheduling and curriculum changes that may result in fee savings.

Another plus is that the fee system provides the department with a reliable and steady source of income to pay for maintenance and replacement. This allows faster turnaround time for needed repairs and ensures the ready availability of small, relatively inexpensive parts such as light bulbs and batteries. In addition, fees collected in one year can be rolled over into subsequent fiscal years, thus allowing the department to accrue a rainy-day fund for large-scale replacements. This has proven to be very helpful in our transition to digital. We were able to plan ahead and accrue funds so that all the camcorders used in the field production course could be replaced simultaneously.

One of the more subtle benefits of the fee system is that it provides a source of department funding that cannot be whisked away or diverted by the upper administration. Since fees represent additional moneys collected over and above tuition, and because they must be spent on the courses collecting them, upper administrators are effectively taken out of the process. Administrators may even find this makes fees an attractive source of revenue, since it means funds can be committed to department needs without the usual political battles and tough funding decisions inherent in the regular budget process. One danger, of course, is that administrators can use fee income as an excuse for cutting a department's regular budget.

At the same time, the fee system can provide a persuasive selling point when requesting new equipment funds from the regular budget. Any equipment purchased for student use with a regular budget allocation can be placed on the equipment depreciation list, supported, and eventually replaced with student fees. This means that every dollar spent on new equipment used in a fee-based course will, over its lifetime, generate more than its own value in fees for maintenance and replacement. If the same amount is allocated to nonfee-supported equipment, it will require even more funds from the operational budget in the form of maintenance and upkeep.

THE STUDENT RESPONSE

While the UNT experience with student fees may not be appropriate for every department or university, it seems likely that many programs could benefit greatly with their own versions of such a system. Yet many institutions are understandably reluctant to implement such a system, since it adds to a student's costs. One academic reported encountering reluctance by administrators at his school to consider fees even for consumable supplies, fearing that the addition of such fees "would be viewed as a way to disguise a tuition increase" (Black, 1998, 132).

At UNT, such fears have proven largely unfounded. The vast majority of our students understand that the fees collected are returned directly back to them in the form of better access to production equipment, fewer out-of-pocket expenses for supplies, and so on. Although no student is happy about having to pay more for education, on balance, the vast majority of RTVF students seem satisfied that the fees are a fair and reasonable cost for a better education in a specialized field they have chosen.

At the same time, RTVF faculty are keenly aware that any additional costs can represent a real burden to students, and they work diligently to keep fees at the lowest possible level commensurate with the level of support needed for a quality program. In addition, the largest fees are charged in elective, rather than required courses, but because students know these advanced courses will provide access to ample, well-maintained equipment, sections invariably fill up, despite the fees.

CONCLUSION

The comprehensive system of student fees used to support production courses taught at UNT has proven to be a major benefit for our program.

Students have access to better equipment that is more efficiently used and better maintained. Faculty members can plan on reliable equipment availability and upgrading as they plan and teach their course. The system's benefits have been especially important as the department has replaced old analog equipment with new digital counterparts. The strong production curriculum has enhanced the reputation of the university among area professionals and resulted in increased student demand and better job placement for graduates.

The circumstances at UNT that led to the system I've described here may be unique. Our state and university accounting rules are flexible, we had an upper administration that was innovative, and our department is of a size and structure that allowed the full benefits of course fees to be realized. Yet I cannot help but believe that with some effort, similar systems could be implemented at many other schools, offering a long-term solution to the chronic problems of equipment funding.

NOTES

1. I should make it clear that the fee system at UNT was in place and operating smoothly before my arrival at the universiy. However, during my four-year tenure as department chair, I came to appreciate the advantages of the system, especially compared to my experiences at other schools where equipment funding and support was a chronic major problem.
2. The students spend the remaining weeks of the semester meeting as a large group in a classroom for orientation, common lectures, and testing.
3. On a larger scale, universities are now confronting the same issue with the replacement of desktop computers. UNT has recently implemented a budget plan that calls for the rolling replacement of one-third of all faculty desktop computers every year. Similar situations at other schools may provide a model for departments wishing to explain to administrators the need for regular replacement of computer-based production equipment.
4. In addition to the fees discussed here, UNT also collects other student fees to cover student health care, technology use (computers and classroom A/V), athletic facilities, and so on. A full-time student taking upper-level courses in some departments can easily pay several hundred dollars in such fees in a given semester. All fees are collected by the university bursar as part of the total semester's bill.
5. Internally, UNT makes an accounting distinction between three separate types of fees collected to support courses: course fees, special service fees, and laboratory fees. The distinction between these fee types is not relevant to the present discussion, and I use the term "course fees" to refer to all three types.
6. Rules are detailed in UNT Policy 2.3.2 "Fees Related to Instruction."
7. The department also has other technical staff paid from regular salary accounts.

8. The one exception is that faculty and staff may use the equipment to prepare for the teaching of the course. An instructor, therefore, may use an audio editing station to install and try out lab exercises and so on.
9. While much RTVF equipment falls neatly into these categories, some items, such as lighting kits and microphones, do not. In such cases, various adjustments have been made. For example, video classes have their own microphones, listed as "video" equipment.
10. The department's Production Committee predates the extensive fee system described here; however, when fees were introduced, it was found that the committee was ideally constituted to recommend how the new income was to be allocated.

REFERENCES

Larson, Mary Strom. (1984). *Choosing equipment and setting policies in radio and television labs.* Paper presented at the meeting of the Speech Communication Association, Chicago, IL. (ERIC Document Reproduction Service No. ED 249 560).

The Student Guide to America's 100 Best College Buys. (1998, 1999, 2000, 2001). Camden, SC: John Culler & Sons.

CHAPTER 4

Teaching Electronic Media: How Much to Change?

Jerry Condra

Today's mass media enjoy the rare opportunity to deliver attractive new products with dramatically improved quality in digital formats. But those same digital technologies have also given birth to a remarkable array of new competitors who threaten to supplant the "old" media. The proliferation of added rivals has led to an unprecedented relaxation of federal broadcast and cable regulation and the consolidation of media ownership. The huge media groups that are now emerging use digital technologies to create enormous savings through economies of scale while increasing revenues through the sale of simultaneous advertising campaigns across many platforms.

Most changes in the media industries have been evolutionary rather than revolutionary. The recent implosion of the dot.com industry seems to indicate that most future changes will take place gradually. John Pavlik (1998) of Columbia University concludes: "Few media technologies are truly new, and even fewer are immediately successful on their initial introduction to the marketplace." He advises media innovators to "watch carefully those who have gone before you and learn from their successes and failures" (p. 75). However, many strategies for success in broadcast management, programming, sales, news, and production—some of them used for more than half a century—appear to be less effective in today's more complex environment. New strategies to fit the changed competition are now being tested by trial and error in daily programming, sales, and management battles, but there has been insufficient time to prove and declare any new media maxims.

In his 2001 convocation address to Syracuse University's Newhouse School of Public Communication, Dean David Rubin (2001) told graduates that during their lives there has been more change in the mass media than in any 20- or 25-year period in history. The fault line for the current instability is the Internet. The media are at a crossroad—bloated, under pressure from Wall Street, desperate to please advertisers, still worried about competition from the Internet. Something's got to give. Your job is to make that something happen.

If students must make that something happen, it is the responsibility of teachers to prepare them for this challenge. As broadcast scholars survey these technology-driven changes radically altering the industry, how do they update their courses when the media marketplace is clearly still in transition?

A survey of 50 broadcasting professors in the United States reveals some disagreement on how much change is warranted and what those changes should be. The interviews were conducted by telephone and e-mail correspondence during June and July 2001. This chapter reports the opinions of those electronic media scholars on how broadcasting courses should respond to changes in the media marketplace created by new technologies.

CURRICULUM CHANGES

Most broadcasting scholars we surveyed feel that an update of their courses should include a revision of course titles. Several point out that the word "broadcasting" historically applies to a terrestrial tower sending an over-the-air signal to a terrestrial receiver. Yet today, 68% of television homes watch at least one set via cable and 15% watch TV from a direct broadcast satellite. Since Internet connections are totally cable or DSL, that leaves radio as the only mass medium totally received by a broadcast signal, and new Internet and satellite competitors are now entering that radio arena.

What new course names are suggested? Most respondents choose "electronic media." Rebecca Ann Lind of the University of Illinois at Chicago chooses "electronic media" because "it is more inclusive, adaptable, and more reflective of the current and future state of things." Tom Streeter, Director of the University of Cincinnati's Streaming Media Project, says, "First, quit calling them 'broadcast courses' unless you intend to communicate that you're preparing students for the wild, wacky workforce of 1973—which way too many programs are, unfortunately. If a course doesn't make any sense in the broad context of communication via all electronic media, get rid of it."

But Doug Underwood of Central Missouri State University disagrees. "I believe that the term broadcast has developed a much broader meaning than just terrestrial transmission. I believe the term 'broadcasting' is perceived today to include all widely delivered media. It's a broad, all-encompassing term now." Michael Keith of Boston University believes "the term 'broadcasting' is kind of archaic. I like the term electronic media as the umbrella term to refer to the industry. But to purge it from the syllabi is premature, since it describes the vital radio industry." Others point out that a change in "broadcasting" would mean inventing other new terms. For example, how would one refer to the "airing" of a "broadcast"?

The issue of comprehensiveness in broadcasting courses extends to the entire curriculum. How inclusive should the coverage of a broadcasting course be in the future? If radio and television have been the electronic mass media in the past, what constitutes the electronic mass media today? Streeter argues that "to teach only about television and radio and claim to have given a student an education in electronic communication is like teaching them about stems and flowers and giving them a degree in biology. Convergence can't be an addition to the holy trinity of TV Production, Radio Production, and Mass Communication Theory. It has to be an integral part of the curriculum—the organizing principle, in fact. The world is networked, digital, and interactive and the curriculum needs to reflect that. Most departments defined their curricula based on the peculiar state of American broadcasting as it existed in 1950. That world died 20 years ago; it's time to bury it."

"Digital and market economies will change the face of TV as we know it," according to Stephen Jacobs of the Rochester Institute of Technology. "It's just going to be a different animal. Personal Video Recorder services like TiVo and Replay are making time-shifting a painless practice and may lead broadcasters to experiment with on-demand offerings for regular shows. The ability to add MPEG 4 interactivity as an overlay to the MPEG-2 digital TV stream takes us beyond the early IP Broadcasting experiments and could make digital TV sets as interactive as any Flash-based Web site or CD-ROM. It's going to look more like a CD-ROM and act a lot more like a CD-ROM. Also, teenagers today demand multitasking, so what is distracting to the older generation is the preferred entertainment mode for the younger generation."

"We're seeing things on the World Wide Web to which broadcast educators must adapt," says Michael Ameigh of Oswego State University. "We must prepare students for a change in thinking from programmer control to viewer control. The compelling aspect of the Internet is that everything is or could be non-linear. If you listen to the plans of Time Warner and others for delivering content, they're tending to think of it as a linear medium. I do not believe that people will choose to watch traditional length TV shows in an Internet environment. The Internet will be used to distribute programs to TV sets or whatever we will use for TV sets in the future."

PROGRAMMING

Susan Tyler Eastman of Indiana University argues that "changes in teaching are needed to make students think that aging faculty are really up-to-date." She advocates changing programming course titles to include either the words "Internet," "Web," or the term "online programming." "Rather than tacking 'the Web' onto the end of course titles," she says, "begin with the online topic or place it in the mid-position." She contends that this emphasis "will contribute to students' perception of you as someone who sees the digital age as the one you are aware of living in, lending credibility when you argue that the past must also be understood." And she predicts another semantic change in how the public refers to networks. "Because we are more likely to refer to 'online services' than to 'online networks,' because vertical integration of media now stretches across all media (horizontal integration?), and because new media exert a backward push on language, we can expect that broadcast and cable networks will soon morph into 'services' instead of 'networks.'"

Douglas Ferguson of College of Charleston, coauthor with Eastman of *Broadcast/Cable/Web Programming: Strategies and Practices* (2002), urges teachers to "get students to look at the communication possibilities rather than the gee-whiz aspects of technologies. How can messages be more effectively communicated? How can producers and advertisers be successful? The hands-on approach to making a Web page may be what students enjoy, but learning Web design for the sake of persuasive communication is what they really need. Give students a framework from which to look at technological change. Teach theory, in other words."

Roger Cooper of Texas Christian University feels that "it's important to give students broad, structural perspective of what programming is and how it functions in all communication mediums. Then, as their career develops, they'll be able to adapt to whatever programming situation they encounter."

Eastman recommends "teaching about current television and Web technology and their inevitable—but delayed—convergence before teaching about the traditional aspects of technology." Eastman suggests that instructors root out all outmoded subject matter in their notes such as "independent" TV stations and scrap them. Eastman urges instructors to remind themselves that "platforms" have become more important than channels, that ownership restrictions have been pretty well eviscerated, that the syndication business is now about foremarkets and aftermarkets, and that producers aim at worldwide popularity as often as American popularity." And she adds: "The advent of the Web has diminished the traditional scheduling strategies in importance, and at the same time, promotion has risen to become a fourth key influence on programming theory." Thus, she suggests adding "a lecture on how Web sites are used by major media to enhance on-air viewing."

Michael Ameigh argues that "there is a huge opportunity here for us to bring our students into a different way of looking at media. This is to see television as nonlinear elements that can be combined into linear format. Take sitcoms: The *Seinfeld* site would have every episode of the sitcom ever produced, and you'd select your favorite episode rather than the one someone else had scheduled for today. For radio and television, this opens up a huge niche market for programming that is easy to access by the user. It will create whole new niche program markets unknown today."

"Programming teachers and students must break with the past," according to Douglas Ferguson. The Internet and broadband technologies have ushered in a new world that is increasingly independent of schedules. Without schedules, you don't need producers—you need content creators. There is a looming paradigm shift, or fundamental discontinuous change at work. The number of TiVo users is less than 1 million, but their 'live' viewing habits are changing in ways much more revolutionary than those brought about by the VCR or remote control. Fewer and fewer media users will be interested in 'live' events, because the line between live and recorded will blur when everything is available 'on demand.'"

Addressing radio, Michael Keith of Boston College believes Internet and satellite radio should be new elements in radio programming courses. "Radio programming is radio programming, whether it comes from a satellite or the Internet or from a terrestrial signal. These new technologies simply expand the meaning of what radio programming is. To me, radio is a generic term. Radio is any programming of an exclusively audio nature."

MANAGEMENT

In teaching broadcast management and sales, Alan Albarran of the University of North Texas says, "Digital represents another way for broadcast managers to distribute content to audiences." He predicts that the immediate impact of digital on management and sales courses will be to add new material describing fresh challenges to broadcast managers. "Those threats," Albarran said, "include new digital and satellite radio competition, covering the costs of new digital TV studio, transmitter, and tower equipment, and substantial costs to market and promote the new digital channel. The final challenge will be to determine whether the TV station will use the new digital spectrum as one HDTV, four to six 'standard' channels similar to today's analog TV, or a hybrid of 'standard' signals during the day and a DTV high-quality mode during prime time. Clearly, no one knows what practices will be successful," he says, " because this is a road nobody has traveled."

"I'm convinced that radio and television are headed for a different business model," says Ameigh. "I believe that radio and TV will be sold on a subscription basis over the Web and on CDs. The cost of broadcasting to 6 or 600,000 people is the same. With digital, the larger the audience, the higher the cost to send out the programs. But the programmer knows exactly what it costs to send a show and can calculate a fair but profitable viewer charge. Commercial broadcasting will continue, but will be increasingly commercial and will operate less in the public interest because it will not be able to afford to. Management and sales classes should begin preparing students for this eventual challenge now."

"The exchange of advertising messages will change dramatically because of the Internet," says Linwood Hagin of North Greenville College. "The flow of information between audience, clients, station, and ancillary support mechanisms has become a digitized channel. Professors need to recognize this fact, uncover new means of information transfers, and prepare students to handle this exchange fluently and efficiently. How we do this is still unclear but will require a quicker adoption of newer technologies by educational administrations and more support for research and exploration of this area linking professionals and educators."

At this point in time, radio has seen more digitizing than television, with local stations employing extensive syndication and local station "voice tracking." Tracking allows local air personalities and those from other markets to feed their part of the show by digital audio. Faced with two national satellite radio competitors and thousands of stations on the Internet, radio executives maintain that their stations will continue to be attractive because of their local coverage and identity.

The *Journal of Radio Studies* conducted a 2001 radio forum featuring radio CEOs and moderated by Ed Shane, a radio consultant and author of *Selling Electronic Media* (1999). Shane followed the forum with advice to broadcast educators for developing radio curricula. He suggested that radio educators emphasize development of local content and also "how to make network or syndicated content *sound* local and responsive, even though it is produced at another location." Noting that "today's media environment does not allow any single traditional medium to stand on its own," the forum recommended that educators devote course time to how programmers can "marry the terrestrial radio station with an Internet 'line extension' to create new audience and sales opportunities."

A second current study targeted to the broadcast educator is "Streaming Into the Future: A Delphi Study of Broadcasters' Attitudes Toward Cyber Radio Stations" (Evans & Smethers 2000). It suggests that "programming, advertising, and promotion course work should include "a greater infusion of marketing and e-commerce principles."

With increasing radio streaming, it urges teaching visual components and Web design, saying, "Future employees will need to understand such principles as typography, color and layout, and Web design." It states that "The peculiar character of Internet news consumers creates the demand that news editors and reporters understand the motivations of these highly critical audience members." Broadcast educators are also urged to specifically focus research and course coverage of "variables that guide consumer's choices among cyber radio stations. For example, which combinations of visual and audio content are most effective at drawing consumers to a particular station's site? Which is more important in dictating station selection, the audio service that is provided or the accompanying textual information? Can you empirically validate the assertion that local content is the chief factor affecting station selection? If stations now have freedom to experiment with innovative formats and programming formats, determine which are the most promising."

Larry Patrick is a former broadcast professor, broadcast executive, and now a station broker. He forecasts that "the Internet, Web-casting, and broadband communications will require students to know how to sell advertising and marketing services to clients who might advertise not only on a radio or television station, but also on a branded news, information, or shopping channel. They must understand and be able to explain to clients how zoned advertising can work for streamed audio or video, and how video production services can customize material offered on a station's Web site." An example of this is a 20-minute video showing an actual laser eye surgery that could be viewed on the station's Web site. The station produced the video, sold spots on the primary channel to promote the Web video, and then delivered prospective customer lists for the eye center based on hits on the station Web site for this demonstration and requests for follow-up material.

BROADCAST JOURNALISM

To Ron Graeff of Oswego State University, students must understand that the new technology has favorable and unfavorable potential. "Digital," he says, "means that anyone with a camera can serve as a reporter, although they might have no journalistic training. Digital video can be altered and changed with few fingerprints to find out who altered it afterward." And he adds, "The security of the media's own information gathering, reporting, and editing systems can be breached. The integrity and authenticity of the system must be protected by a tight security system." The positives, however, he identifies as "huge new data bases and powerful search engines that give new journalists access both to raw data and edited materials at lightning speed. Classes should include instruction in computer assisted reporting. Finding and verifying information at high speed and in huge quantities can be challenging, but rewarding."

Despite the advances in technology, most educators agree that careful attention must be paid to helping students develop skills of analysis and interpretation. "You can't ignore technological change, but you can't be a slave to it, either," says Robert Papper of Ball State University. "Nor are we operating trade schools. I attempt to teach students how to analyze and process information, how to reason, and how to tell stories using pictures and words. Notice that technology isn't part of it. I do tell students that they're in the difficult situation of having to learn both analog and digital because that's the world they're entering. But storytelling techniques are largely unchanged; the rules of composition haven't changed."

And Barbara Croll Fought of Syracuse University agrees. "A computer should never be an excuse for not doing original reporting. Certainly it provides great resources for reporters to background themselves. Then, they should still get out and meet the people they're covering face-to-face." Regarding the inaccuracy of much data found on the Internet and e-mailed to newsrooms, she believes that "we have to teach students how to evaluate material correctly with independent sources. But this isn't new. Reporters had to deal with impostors and inaccurate information long before the Internet. Now, we must train students in copyright and ethical issues and address attribution and sourcing with special care."

"Interactivity is the killer application of digital technology," says Graeff. "It allows the news source to interact with the audience. Future viewers will be able to select only the stories in which they are interested and download them. Some viewers will no longer watch a pre-edited report of what an editor thinks are the top stories of the day. Instead, they will choose stories they want to see from a news menu. Students must understand how to cover a story such as a press conference and produce user options to watch one or several short packages on major angles of the story, watch the entire news conference, and see graphics and documents from the press conference."

Randall Wenner, Broadcast Journalism Laboratory Manager at Syracuse University, says students there have been doing all radio news editing digitally for more than five years. Students take analog recorders into the field, but digitize the tape inside the computer. "We're focusing on use of the computer as an editing tool," he says, "but we're afraid that if we teach students only digital, they will show up on their first job and find analog editors there. We're trying to mirror the industry by teaching both analog and digital video editing." Most faculty interviewed for this survey shared this approach. Wenner said, "We have also redesigned our introductory courses to introduce students to all media that might be parties to convergence, so that if they must rewrite their TV story for the Internet, it won't come as a complete surprise. Use of supporting materials on the station Web site is also required."

"We've made no changes to our journalism courses because of digital per se," says Papper. "On the other hand, we are making significant changes because of the Internet. Unlike digital (which is just a better way to manipulate information and images), the Internet poses the challenge of a new medium—a blending (in our view) of print and broadcast. And we're responding by doing just that. Effective in fall 2002, many of our print journalism and broadcast journalism courses will merge and include a component directly linked to the Internet."

PRODUCTION

"Change has arrived in media production," says Robert Musburger of the University of Houston. "But we must look at the changes with an accurate eye, not one of an emotional heart. We must teach our students the differences as well as the similarities between analog and digital. Analog will be with us for some time and we cannot forget that all human communication is analog: We see, hear, speak, and respond to light in an analog mode." Musburger urges production instructors to learn the digital technologies, "but they should not forget the analog portions of the process. The two most important areas are in non-linear editing and creation of graphics materials. We and our students have become infatuated with digital and tend to ignore the acquisition of signals by concentrating on the digital processing aspects. A balance must be reached for a complete education in media production."

But Lynne Gross of California State University at Fullerton says, "I don't agree with that. Students are coming in today with non-linear brains. It's part of the attention span thing. I'm not sure if it's better, but they flit around from topic to topic as we do on the Internet. I don't think that they look at filmmaking as a linear experience. They shoot the middle one day, cutaways the next, and the opening on the last shoot. But movies have always been made that way. Perhaps distribution is still linear, but I don't think that shooting and editing are."

Ameigh quotes digital film pioneer George Lucas as saying, "When I go to USC's Annenberg School of Film, the students aren't using the language of film anymore, they're talking about capturing sequences of video and then talking about how to put them in some coherent whole. They've got their DVCams. They don't worry about lighting. They don't worry about the same kinds of conceptualization that go into story design. They want to be able to go out and shoot something and see it as an element of a potential project that will come along some day and they'll know they have it in their back pocket."

In fact, "digital video cameras and editing programs have been adopted at film departments across the country" (Carlson 2001). "Those who teach filmmaking say that digital video technology has allowed students to experiment with different editing styles, present work to the class more quickly, and to focus on the aesthetics of films, rather than on the technical details of how to put them together. 'It's definitely a revolution, and it's here,' says Julie Corman, chair of the graduate film program at N.Y.U., which recently purchased a dozen Apple computers for an editing room. One student comments, 'Major filmmakers haven't fully explored the digital medium, so it offers us an edge. By the time we graduate we'll have more digital video experience than most filmmakers.'"

"Are we really at a paradigm shift?" asks Lynn Spangler of the State University of New York's College at New Palz. "Or is it the same thing we've been doing for years in production, and that is 'We have new technology and new ways to make cool stuff.' It's being distributed in different ways, but is it really different when it's on a computer? If you look at the advice on how to shoot for video streaming, does it ever sound old fashioned? 'Don't move very much. Keep a simple background. It's like we're going back to *Howdy Doody Time*.'"

To Norm Medoff of Northern Arizona University, "It makes no difference whether it's digital, analog, or mechanical as far as I'm concerned. A camera is a camera. But putting together images that make sense, writing, and editing are almost totally unchanged by digital. One thing that does change is the ability to put all those things together easily in post production. Three-fourths of the students will never be editing professionally. I don't worship post production—I worship videography."

Peter Orlik of Central Michigan University cautions "It is important that we do not allow focus on new applications to replace teaching of fundamental insights and competencies. Technologies constantly broaden and improve. But some core principles and skill sets remain the educational underpinning for any maturing media professional." Orlik advocates a "rigorous set of writing experiences" and suggests that students "be taught procedures for thorough critiquing of their products and those of their competitors because a true professional is able to systematically and continuously appraise the performance of its practitioners. Certainly we need to modify units of existing courses, add new digital production experiences, and delete topics or courses that digital technology has rendered obsolete. The bedrock competencies necessary for success in our enterprise will not change, however, and the teaching of them will continue to separate professional from trade school pedagogy."

Gross agrees: "We should spend more time with aesthetics. But you won't do that by reading a book. You must have adequate equipment to, say, work with music that perfectly backgrounds a video production." And Doug Underwood adds: "Telling the story is the most important thing and equipment should be relevant to that. But the type of equipment used is relevant because we are not graduating people to work with consumer equipment. They will be working with professional equipment. Graduates with the ability to operate professional equipment are the first to be hired. And Lynn Spangler asks: "Do we teach them the old and the new technology at the same time? Our students may be working where they must do linear editing. We must also bridge the analog and digital world for another reason: Financially, you don't throw out your good stuff just because there's something new on the block."

Vince Doody of Oswego State University points out that "in most colleges, budgets mean that our conversion must be done bit by bit. But that's also how the broadcasters are moving to digital. Many universities do not seem to plan conversion on a long-term basis due to the uncertainty of annual funding." Doody cautions educators to think carefully before accepting gifts of used equipment unless they have technicians to install and maintain any donated equipment.

Medoff believes courses must emphasize the skills that will be most in demand in radio. He says that "while some entry level radio jobs are declining, our students are hopefully learning how to design Web sites, stream video, interact with the audience in a very direct way, and design sticky Web sites that keep the audience coming back. I'm seeing some students leave here with entry-level Webmaster jobs that pay $50,000. Our courses must address to those kinds of skills." And Michael Keith asserts, "Radio jobs are changing. They aren't going away. We must educate our students about which radio jobs are going away and prepare them for the jobs with potential. Radio groups will be looking for streaming program directors, radio/Internet sales representatives, and promotion managers."

CHANGING TEACHING METHODS

Programming

Ferguson advises instructors to "get students thinking about the difference between old-style lean-back entertainment and new-style lean-forward media use. Study the functions of entertainment and information programming. Then list the functions on the chalkboard and contrast new possibilities with the status quo. Rather than examine the technologies too closely, look at the new issues that result from technology change: economics, privacy, copyright, etc. Learning comes from students thinking hard about what lies beneath the surface of technological change."

And when using technology in the classroom, Ferguson believes that "less is more. Use videos but limit their intrusion into class discussions, wherein the real meaning takes place. If necessary, put videos on closed reserve in the library. Likewise, recognize the dangers of the overuse of PowerPoint. Is the presenter engaging the listeners, or forcing a one-interpretation worldview on them? Don't let PowerPoint become a conversation killer. Again, the goal is not training or indoctrination. The goal is education and learning."

Management

Patrick contends that, in the new multichannel environment, college sales classes must train the future broadcast salesperson or manager to understand the importance of measuring audiences by traditional Nielsen,

Arbitron, or Web-based hit-counters. But, he states, "counting hits alone is not enough. Advertisers will need to know how the multi-channel marketers can prove to them that they are driving consumers not just to look at or hear the material but help track respondent purchases and consumer selections."

Albarran suggests that the digital and regulatory challenges to broadcast management be tied directly to learning activities. Students could intern on a fund raising campaign for a public TV station's digital conversion, research the cost of converting local commercial TV stations to digital, research the cost of marketing the new digital channels, and interview local radio station managers on how they plan to compete with the new satellite and Internet radio competition. Sylvia Chan-Olmsted suggests "incorporating more 'real industry/company' case studies as the Internet is providing extensive background information for in-depth analysis." She offers an example from the University of Florida Library (2001). Gross also uses "at least one case per three hour class, but I don't get students to do exercises on issues such as how managers solve the problem of HDTV cost, because by the time they are managers, the problem will have been solved for a long time." She adds, "I have tied some management exercises directly to the campus radio station."

"We should be teaching the reality of the radio industry—that of consolidation, convergence, and content" says Sam Sauls. "As radio stations move toward automation and more satellite-fed programming, the radio professional will be concentrating on station sales and promotions. This is where the student needs to understand the overall business of the radio industry. We must incorporate these concepts into our classes. Industry professionals should be brought into the classroom to present first hand knowledge."

Judie Thorpe of the University of Wisconsin at Oshkosh recommends extending digital to class exercises as much as possible. Her students' proposals to advertisers are computer produced, and each student is required to design a Web page for their "clients." Lionel Grady of Southern Utah University says, "I could see the usefulness of incorporating one of the software accounting programs to make the students more aware of the financial aspects of management and sales. There are also excellent software programs that digest TV ratings and cost-per-point as a TV sales tool. I know most TV sales departments use TV Scan, but I'm not aware of any college professors using it, possibly because of the prohibitive cost."

William Covington Jr. of Edinboro University in Pennsylvania urges educators in the midst of this digital evolution not to lose sight of "the intangibles such as working relationships and the expertise of veterans in the industry. These forces will have a direct positive impact on a media company's operation. Digitization will require learning to take place at all levels in a media organization. This learning will transform individuals as the organization grows and changes. A significant amount of attention should be paid to the intangible human elements influencing these changes in the industry."

Learning activities in broadcasting sales classes need added material, according to Patrick. "Clearly students need to go far beyond just doing a customer needs survey leading to an attempt to teach them how to sell spots," he says. "They need to understand the entire marketing focus of a company and how best to expose customers to the products and services of clients." Patrick cites the ability of large broadcast groups such as Clear Channel to tie together radio, television, outdoor, direct mail, ticketing, and concert venues.

Broadcast Journalism

"Computer-assisted reporting allows students to get direct access to mountains of data such as government reports," reports Graeff. "For instance, a student reporter could not only research the performance of a real local prosecutor, but also compare that prosecutor's performance with other real prosecutors across the state. Students with this skill are better prepared to recognize and report more sophisticated stories when they begin entry level reporting."

Barbara Croll Fought adds: "There are great opportunities for us as faculty to integrate more digital material into our exercises. Instead of printed handout exercises, our students can now access real stories on the Internet to use in writing exercises. I think this gives real currency and excitement to teaching journalism." Graeff adds: "AP NewsDesk and other software allow students to use the same tools that newsrooms use." Ameigh believes that many opportunities to make course materials interesting to students are overlooked. He cites foreign language classes that use popular DVD movies with various language tracks for instruction.

Randall Wenner reports genuine student interest in accessing streamed news broadcasts from major market stations and analyzing them in class. And "in using PowerPoint, I make printed handouts to accompany what appeared on the screen. The student's feedback is 'I loved the handouts; they helped me focus on the discussion at hand and not so much on note taking.'" Other issues involving teaching methods mentioned in this survey were the in-class production of writing, video, and student news broadcasts that could also be used for student portfolios and capstone experiences.

One common complaint of television news directors is the lack of qualified producers. Naughton (2001) studied this shortage in the Central New York area and determined that one of the reasons broadcast journalism graduates were disinterested in producer positions was lack of knowledge about what producers do. He concluded that broadcast journalism courses should devote more class time to the role of producers and include exercises in producing, in light of the many entry-level positions available in this field.

Since some broadcast employers are now requesting "content producers" capable of working simultaneously in print, broadcast, and Internet, much thought is being given to how to educate students for this role. Sylvia Chan-Olmsted of the University of Florida reports, "We are seeing more integration between print journalism and broadcast journalism courses. Students from these two departments are actually working together and getting to know each other's discipline better."

As to training students specifically for Web news producing, Croll Fought feels "it calls for thinking more multi-dimensionally in terms of packaging audio, video, graphics, and text. It's a different frame of reference from those who came up in the broadcast culture of moving images. It takes different skills to package and parse the information in ways that are interesting to the viewer." She is also hopeful that "once we get over this digital learning curve—which is similar to moving from film to video—I think we, as faculty, may need to spend less time on technology and spend more time on important things like content. We should be able to expect students to do more and know more, because they have more content available to them and we can expect higher standards technically. They need to think in terms of packaging audio, video, graphics, and text in one story, all edited on one computer."

Production

Lynn Gross believes that "digital has already changed production classes by speeding up editing and making it easy to do special effects." Because students grew up using computers, it facilitates teaching the concepts. Digital allows more time to deal with aesthetics.

Streeter suggests exercises to match the available technology. "A good edit is a good edit, whether it's cut on film stock, a straight cuts edit system, an Avid, or two stacks of photo flipbooks. Telling yourself that it's good for them because they're using a computer is ridiculous. They'll use a computer when they're doing the inventory at the local Piggly Wiggly, and that's all you're getting them ready for. If you only have the equipment to teach flipbook editing, do that—and do it well. Don't try to do more than you're capable of."

Karen Kearns of California State University, Northridge, claims, however, that "lab assignments must be digital and computer based. They must work on DAWs and work quickly. Except in tiny markets, that's what everyone is using. I still believe students in performance classes need experience in live presentation. It will help them whether they are taped and edited or not. I still have deejay type board tests and newscasts. Promotion is the key to audience building, but I don't know if the class is the best forum for teaching promotion. I encourage my students to intern in promotion at local stations. Other ways to teach is for students to create a promotion campaign for a show or station or having students analyze promotions running on local stations. And the ultimate learning tool is the campus radio station, which should be run using commercial radio format standards with a slight leeway in play list. It teaches the discipline of following a format 24/7."

Digital has introduced a new challenge in production labs, according to Doody. "There's a storage problem with work in progress by students. Unless you have a massive server or each student buys their own digital storage device, there is not room to keep every single piece of work by every student through the semester. Students must work intensively on production, complete the project, and release the space. That makes it difficult to do long form productions where students work throughout the semester on a project."

In radio production, what teaching method works best? Sauls (2000) reports that "though students can use a computer, that doesn't make them production literate. The concepts of audio (sound) and radio production (mixing, layering, transitioning, etc.) are best learned—for whatever reason—in the analog world. Some of us believe that because the student is sitting in front of an audio console and working with physical tape and equipment, a more conducive atmosphere to production exists. If a student just learns a 'system,' that student will be able to work only on that specific model. But if the student can understand the theories of digital audio, these can be applied to all systems."

In TV production, how can instructors move students toward a more multimedia approach to production? Step one, according to Stephen Jacobs, is to "recognize that your students probably are more familiar with computer technologies and the Web than you are. Start getting yourself educated with some of the simple packages that allow you to combine video with other media and control them. Director is the best choice, but it has quite a learning curve. Flash is more popular on the Web but doesn't handle video quite as well as Director. iShell and Livestage are other alternatives. Read *AV Video* and *Multimedia Producer,* take a look at Communication Arts Web site and Interactive Design annual issues, streaming video Web sites and everything else you can get your hands on, and streaming video magazines. It's now enough to have the camera shooting a TV show cinema verité style. You must think and plan in advance. In their early days as CD-ROM producers, the Discovery Channel determined that they needed to shoot 40% more footage in an original production if they wanted to make a companion CD-ROM. It's likely that a similar metric will crystallize as we evolve to make television more interactive. You have to plan to shoot many additional perspectives and issues. You must look at every story or production you'll do with an eye to how can I help people experience this story beyond what I can show? How can I leverage my assets to work well? How can I tie this into my Web site to make people want to see more?"

CHANGING EQUIPMENT AND FACILITIES

Many experiments with streaming audio and video are now being conducted by mass communication departments, but quality limitations continue to restrict use of the new process. Web traffic can slow down or stop video while the audio continues ahead of it. It is clear that increased streaming awaits improved bandwidth (Copeland, 2000). Typical of mass communication tests of streaming is Bowling Green State University's use of Web video as a companion to a student-operated TV news program (Nordstrom, 2000). For class instruction and exercises, Michael Ameigh's feeling that "it's easier to send students a video, DVD, or CD with the same data on it" seems to prevail.

Broadcast Journalism

"The bigger technology problem may come from a mixture of student expectation and competition," says Bob Papper. "More and more, we're seeing students who have been introduced to digital technologies in high school, and they're shopping for certain toys as they look at colleges. I know we're feeling the pressure, and I'm sure other schools are, too." Most of the respondents in this survey faced the same problem.

Graeff predicts that "distance learning can be used effectively to teach broadcast journalism. The future may see the addition of videophone connections. Where the student's computer is not equipped with video editing software, online services allow students to shoot video, download it to a Web site, edit the story by remote commands from the student's computer, and the professor can access the Web site to critique and grade the video."

Production

Musburger does not hesitate to address a problem familiar to many mass communication departments head on. "We are finding more and more students arriving at our doors with nearly as much training in digital production as we offer, which they gained at an advanced high school or a community college. This will continue as more and more lower level schools offer video production, often taught by faculty with little or no professional or theoretical experience. This may mean that, in the future, upper level universities and colleges will need to concentrate their efforts on teaching the theory and ethics of production, writing for media, budgeting, and the finer points of professional producing. We will continue to be swamped with individuals who want to learn how to operate their digital cameras in a week or two. Some students will expect the same. I believe these problems are as important as the difference between teaching analog or digital media production."

This view of this role for four-year institutions may bring comfort to many administrators in colleges not well endowed with new digital equipment. They wonder if students used to cutting-edge technology in a community college will find their dated production facilities attractive enough to lure them in for the final two years. They also question which students will appreciate more—high-quality instruction in the finer points of producing or top-of-the-line toys to play with.

Doody recalls that community colleges have historically been more responsive to adopting new equipment, and four-year institutions must keep up. He suggests asking yourself how students think. "Put yourself in the place of students who have spent two years in a community college with really good equipment. If they visit a four-year institution where they might complete their degree, but find equipment that's not up to date, they view that institution as not being committed to production and, therefore, not committed to them. That's when they decide to attend another school."

Ameigh has an entirely different take on the matter of production equipment. "We're thinking in the old box that sees equipment as a showcase for the department to lure in new students. That model is dated and doomed. Most television program production is done on the fly, without lights, and using very inexpensive equipment. The new sophisticated abilities of digital broadcasting and production equipment at a huge decrease in price has changed everything. A laptop computer and $60 in audio editing software will do almost everything a radio studio can do. Instead of paying $250,000 for a new radio studio, institutions today should furnish students with a laptop. The key issue is teaching students how to make great shows and spots. We should prepare students to think in a media context—not to run a certain piece of equipment. Should we teach to the 'broadcast standard'? The broadcast standard has changed."

Spangler agrees with Ameigh, and observes, "Think about the whole notion of the video Toaster. It was to put expensive production capabilities on workstations and home computers. Remember the survey *Training Tomorrow's Broadcasters: A Comparison of Employers' Needs and Educators' Intention* (Lind 1999). They really want us to do on-the-job training and that's not our job. We're not trade schools. But we still need to find that happy balance where they're getting theory and we're focusing on content and aesthetics, but they're also learning and feeling comfortable with equipment."

"That's true," comments Lynne Gross, "but a laptop and some software aren't going to get them a job editing with Avid in Hollywood." We're using Final Cut Pro and I-movies software in our classes and they work well for beginning students at home. But professional employers demand a higher level of sophistication."

"A computer is current for two years max," adds Streeter. "We've been communicating for 5,000 years. Do the math. Buy only as much equipment as you need to teach the lessons you want to teach. Make sure there really is a clear set of lessons you want to teach. But don't get attached. In two years there will be new equipment to buy, but your lessons should survive. If they don't, you were teaching the wrong thing anyway."

CONCLUSION

The faculty members interviewed were unanimous in their belief that teaching core principles, insights, skill sets, and concepts is the most important thing they do. They strongly recognize that knowing how to operate technical equipment does not qualify students to use that equipment to produce memorable and evocative content. They also emphasize that equipment evolves continually, while the principles of producing media content change much less frequently.

These also believe that the "broadcasting" field is being slowly transformed into a new multi-platform "electronic media" marketplace where technological preparation and skills will be mandatory for success. Some of those essential skills are not now required for many "broadcasting" degrees or even taught in traditional mass communication departments.

Most professors surveyed believe that the "broadcast" curriculum must be renamed and revised comprehensively to include the new electronic platforms, that faculty must be competent in the technology areas of the courses they teach, and that new teaching methods and equipment must be available to effectively train students in this wider new field. They point out that this might be achieved through creating interdisciplinary programs with other departments such as computer sciences and graphic arts. But they insist that—however it is done—these significant changes must be made to adequately prepare our students for this new media world.

What does the future hold? Will broadcast faculties formulate an updated curriculum and argue their case persuasively? Will administrators recognize and act on the need to modernize equipment and broaden the curricula in response to a radically changed electronic media? Will academic turf wars delay or kill such efforts? This is one drama that won't be broadcast. It will play out in mass communication departments everywhere.

REFERENCES

Carlson, S. (2001, June 22). College Filmmaking Programs Get a New Look With Digital Video. *Chronicle of Higher Education*, A31.

Copeland, L. (2000, May 22). Digital Video. *Computerworld*, 34, 83.

Eastman, S., & Ferguson, D. (2002). *Broadcast/Cable/Web Programming: Strategies and Practices*, 6th ed. Belmont, CA: Wadsworth.

Evans, C., & Smethers, S. (2001). Streaming Into the Future: A Delphi Study of Broadcasters' Attitudes Toward Cyber Radio Stations. *Journal of Radio Studies, 8*, 1. pp. 4, 25, 26.

Lind, R., & Danowski, J. (1998). *Training Tomorrow's Broadcasters: A Comparison of Employers' Needs and Educators' Intentions*. Presented to the National Association of Broadcasters and the Broadcast Education Association April, 1999. http://tigger.uic.edu/~rebecca/educexec.html

Naughton, P. (2001). Where Have All the Producers Gone? Senior Honors Thesis, Oswego State University.

Nordstrom, M. (2000). Posting Web Video as a Companion to a Student-Operated TV News Program: A Case Study. *Feedback, 41, 4*. pp. 1–7.

Pavlik, J. (1998). *New Media Technology: Cultural and Commercial Perspectives*, pp. 75–76. Needham Heights, MA: Allyn & Bacon.

Rubin, David M. (2001, May 20). Rapid Change Poses Challenges for Communications Grads. *Syracuse Herald American*, p. D-3.

Sauls, S. (2000). *The Culture of American College Radio*, pp. 162–163. Ames: Iowa State University Press.

Shane, E. (1999). *Selling Electronic Media*. Woburn, MA: Butterworth-Heinemann.

University of Florida Library (2001). Dell Computer Case. http://Web.uflib.ufl.edu/cm/business/cases/dell_computer.htm

CHAPTER 5

Technology and Academic Collaboration: An Interdisciplinary Model

Gerard Donnelly

In writing about interdisciplinary research and scholarship, Wissoker (2000) asks, "Is this 'inter' a bridge connecting two ways of working? Or is it some third way, one that is beyond them?" The question also can be applied in the larger context of interdisciplinary academic programs. Is it possible, through collaboration, to do a better job of preparing college students for an increasingly complex world by forging connections between departmental perspectives? Is it advisable to create entirely new programs of study whose focus goes beyond what any one academic department would offer individually?

The transition to digital technology has brought with it an opportunity for Mass Communication programs to collaborate with academic departments such as computer science, art, and music—departments that not long ago had little if anything in common with the mass media. With the development of the Internet and because these departments utilize similar software, they suddenly have a great deal in common. This chapter will trace the birth of a new interdisciplinary major known as Interactive Digital Media at Northwest Missouri State University. Although specific circumstances always vary from one campus to another, it is hoped that this interdisciplinary initiative, and the manner in which it came about, might serve as a model for faculty and university administrators to consider when planning new academic programs in response to changing industry needs.

The Interactive Digital Media major includes core classes in Mass Communication, Computer Science, and Art. Students also choose a concentration from one of the three departments: new media, computer science programming, or visual imaging. The major grew out of a relatively new multimedia minor that included courses in Mass Communication and Computer Science. The IDM major, as it came to be referred to, was in the planning stages for several years and won approval from the Missouri Coordinating Board for Higher Education in 2000.

Among the topics covered in this chapter are the industry changes that proved to be a driving force behind the new major, the collaboration involved in designing a curriculum, the importance of generating support by aligning with university initiatives, and the human elements that proved essential to successful planning and implementation.

RELEVANCE TO INDUSTRY

According to Northwest's official application to start the IDM major, the purpose of the new program is "to produce graduates who are proficient in written and visual communication/design, programming, and legal and ethical considerations as they relate to the Internet, the World Wide Web, and electronic digital media." The program was a response to the development of modern industries such as computer graphics and animation, interactive education and training, and electronic entertainment, as well as advances in electronic journalism and electronic advertising.

The proposal goes on to state that the growth of the Internet, World Wide Web, and new digital media has created a demand for skilled employees in areas such as electronic graphic designer, graphic production artist, CD/DVD interactive digital media producer, interactive educational producer, interactive entertainment producer, Web editor, Web content producer, Web master, Web developer, Web programmer, and electronic media journalist. The point is made that to succeed in these areas, an employee needs a combination of creative abilities, communication skills, and technical expertise.

A committee of faculty from the three departments conducted research to determine the level of demand for these positions. Craig Warner, an Associate Professor of Art with extensive professional experience in the graphic arts industry, took the lead in contacting industry representatives in the region. "I kind of knew what to expect," Warner said. "I knew that out in California there was a group of industry leaders, mainly in the entertainment industry, who kept asking why they couldn't find computer people with an aesthetic bone in their bodies. They were complaining that universities send them these computer jockeys who can't paint, they can't draw, they have no composition sensitivity, and then they're expected to sit down and do major motion pictures and major games. Or just the opposite is true: the industry gets people who can paint and sculpt and draw and do all these things and when asked to help with the network or load something or de-bug something they get this blank look on their faces" (Warner, personal communication).

Although phrased more subtly, industry professionals in the Midwest offered similar assessments. Warner sought ideas from agencies and commercial studios in Kansas City, Minneapolis, Des Moines, and Omaha. "Most were bemoaning the fact that their new employees just weren't prepared, they weren't versatile. I asked them if they seem as bright as previous people and they said yes there is no problem there, it's just that they're not well-rounded." Warner noted that the problem was much more intense for small and medium-sized shops such as Fearless Eye in Kansas City, where employees were expected to wear more hats. At a large ad agency such as Fallon McElliott in Minneapolis, there were enough specialized positions that the lack of versatility was less of a concern.

The initial research included a brief questionnaire sent to working professionals in the multimedia industry. The following questions were included:

1. What is your reaction to the name of the major, Interactive Media? Do you have any other suggestions?
2. Are there enough programming classes in the core?
3. On a scale of 1–5 (with 5 being very important) please address the importance of each class listed below.
   - 3a. What about an "Issues" course offered each year that would examine industry and professional development, as well as examine and evaluate the latest multimedia software?
   - 3b. How important is it to have a media management class in the core?
   - 3c. How important is it to have a management information systems class in the core?
   - 3d. How important is it to have a marketing class in the core?
4. Should there be a marketing concentration in the beginning or should we keep focused with the developer-type courses?
5. Are there any courses you see missing?
6. Regarding the financial commitment from the university, how often would we need to replenish computer hardware and software? Life cycle for updated equipment/software?

Although the number of responses was less than hoped for, the survey did elicit some useful opinions from industry professionals. Specifically, those who responded were enthusiastic about the name of the new major, were pleased with the commitment to programming courses, were divided about the wisdom of including more management and marketing courses, and stressed the importance of updating equipment every two to three years and software "whenever there is a quantum leap."

Additional research included an online search of 14 national newspapers, which found a total of more than 6,000 listings for interactive media jobs in two days. Also striking were the salaries for these new media workers. The average compensation ranged from $46,734 for a design/layout position to $80,750 for a project manager. Average pay for a Webmaster or Web designer was in the $50,000 range.

Each of the three departments already were offering courses that provided relevant skills for the IDM major, such as visual aesthetics, communication, technological knowledge, computer skills, animation, and storytelling. However, no one department covered all these areas or possessed anywhere near the resources necessary to offer the major by itself. Further, the committee's research found that there was no similar program at an institute of higher education in Missouri and that there was only one in a neighboring state. The research included comments from an instructor at an area technical school who expressed disappointment that his multimedia graduates had no choice but to attend a school on the East or West Coast if they intend to pursue study at the university level.

To document the current level of interest among students on campus, a survey was conducted of students in the Freshman Seminar classes for Mass Communication, Computer Science, and Art. Out of 152 freshmen students surveyed, 46 indicated that they would be interested in majoring in Interactive Digital Media. Considering the size of the university, which enrolls approximately 6,000 students, and the size of the individual departments, which range from 100 to 250 majors, the figures were considered very encouraging.

CURRICULUM DESIGN

A committee of faculty members representing the three departments became the driving force behind the design of the new major. That design underwent several changes during the time that it took to gain all necessary approvals at the campus level from the department faculty, the chairs, the deans, the faculty senate, the provost, and the president. The first change occurred when the committee decided there should be three different concentrations rather than having the same exact course requirements. Whether a student chose the 21-credit-hour concentration in New Media, Computer Science Programming, or Visual Imaging, they would also need to meet common core requirements totaling 40 credit hours (see Figure 5–1). The total 61-credit-hour program qualified as a comprehensive major in which no minor would be required.

The courses in the core provide students with an understanding of the industry environment and with most of the skills they would need regardless of which specific field they entered after earning the degree. These courses include Design, Photography, Creative Electronic Imaging, Letterform and Graphic Design, Writing for Media Professionals, Media Design, Multimedia Production, Communication Law, Web Publishing, Visual Basic, Multimedia Development, Scripting Programming, Marketing, and Senior Seminar. Marketing was the only course from a department other than the three to be included in the curriculum. The objective was to include e-commerce in the program and also to open the door to the possibility that a fourth core in this area might emerge in the future.

Interactive Digital Media Major
61 Credit Hours
Art Computer Science/Information Systems Mass Communication

Interactive media workers need to possess cross-disciplinary knowledge and skills in areas that deal with visual aesthetic understanding and communication, integrated with technological knowledge, in order to be effective and efficient employees. The goal of this new major is to produce graduates who are proficient in written and visual communication/design, computer programming, and legal and ethical considerations as they relate to the Internet, World Wide Web, and electronic digital media.

CORE (40 credit hours)

13-191 Introduction to Design (3)
13-201 Creative Photography (3)
13-292 Introduction to Creative Electronic Imaging (3)
13-294 Letterforms and Graphic Design (3)
20-111 Intro. to Broadcast Operations (3)
20-243 Media Design I (3)
20-303 Multimedia Production (3)
20-314 Communication Law (3)
20-403 Web Publishing (3)
44-140 Introduction to Visual Basic (3)
44-333 Multimedia Development (3)
44-335 Script Programming* (3)
Senior Seminar: 13-402 or 20-416 (1)
55-330 Principles of Marketing (3)

*New Course

CONCENTRATIONS (21 Credit Hours)

New Media
20-130 Writing for Media Professionals (3)
20-150 Journalism Practicum (Yearbook) (1)
20-151 Journalism Practicum (Newspaper) (1)
20-225 Television Production (3)
20-330 Advanced Television Production (3)
20-343 Media Design II (3)
Take one of the following courses:
20-350 Adv. Journalism Practicum (Yearbook) (1)
20-351 Adv. Journalism Practicum (Newspaper) (1)
20-354 Magazine Practicum (1)
20-353 Principles of Print Advertising (3)
20-419 Special Projects in Broadcast Internship (3) or 20-452 Journalism Internship (3)

(*continues*)

Figure 5–1. Curriculum for Interactive Digital Media Major

Computer Science Programming
44-131 Introduction to Structured Programming (3)
44-230 Fundamentals of Computer Science (3)
44-235 Data Structures (3)
44-345 Computer Organization I (3)
44-455 Networking (3)
44-460 Database Systems (3)
44-300 or above CS Level Elective (by advisement): 3

Visual Imaging
13-120 Drawing (3)
13-221 Life Drawing (3)
13-240 Painting (3)
13-394 Advanced Creative Electronic Imaging (3)
13-497 Creative Digital Animation (3)
13-300 Level Elective (by advisement): (3)
13-592 Graphic Design Internship (3)

Figure 5–1 (*continued*). Curriculum for Interactive Digital Media Major

Each department contributed approximately the same number of courses to the core. However, at the outset the committee members agreed that they would choose the courses best suited to accomplish the goals of the new major and that it was not a priority for each department to provide the same number of courses. Students who chose a concentration in New Media would gain additional experience in Mass Communication classes that cover areas such as broadcast operations, radio and television production, media design, advertising, and a professional internship. Students in Computer Science Programming would learn more about data structures, computer organization, networking, database systems, and programming. Students in Visual Imaging would take Art classes that taught drawing, life drawing, painting, creative electronic imaging, and digital animation, along with a professional internship in graphic design.

TEAMWORK

To identify the elements that went into successfully launching this new major, in-depth interviews were conducted with members of the faculty instrumental in designing the curriculum. All emphasized that one key was the ability to work as a team and minimize territorialism. That might sound rational enough, but at colleges or universities with entrenched faculty, obstinate perspectives, and conflicting priorities, it often becomes a Herculean challenge. Originally, the idea for the program emerged through a series of conversations between Jodell Strauch, Assistant Professor in Mass Communication, and Carol Spradling, Assistant Professor in Computer Science. From her participation at national conferences,

Strauch had noticed how the two fields had grown so much closer together in recent years. As interest in computer-assisted reporting grew to include building Web pages, and as use of the Internet seemingly exploded, more computer science specialists were getting involved with journalism and more journalists were learning about computers and programming. Almost all of them were making the adjustments on the job as working professionals. It seemed like a good idea to design an academic program that would give students a head start in preparing for the realities of the new workplace.

Strauch explained that the group built the major around the multimedia minor already in place, which itself had grown out of a recognized industry. Again, minimizing territorialism was a key: "We thought about how to create the perfect major without thinking about which classes exist and who teaches what. We decided we needed creativity, we needed art, we needed graphic design, we needed illustration, we needed programming. So we came up with a list of the skills needed and didn't think about where those skills were located, whether in this department or in that department. As we went along we found those skills centered on mass communication, computer science, and art" (Strauch, personal communication). Finding the right mix was not easy though, because different entities in the industry were looking for different characteristics. "My research into job postings showed that Web developers and programmers needed to know specific software such as DreamWeaver and Quark while designers needed some software skills also but not as much," said Strauch. "It was more important for the designers to be artistic and creative."

Strauch and Spradling both agreed that honesty and a lot of give-and-take were crucial ingredients in those early meetings when core requirements were being decided. Strauch credits most of the group's success to the personalities of the individuals and their willingness to work together: "The one thing that will make or break an interdisciplinary major is personalities. If any one of us was more obstinate it would not have worked. I would say it was eighty percent personalities and twenty percent hammering out the right courses to design the program." According to Spradling, "The fact that most of us already had a relationship because of the minor helped. There were times when various departments had to give things up. We had more hours in the beginning than we were allowed and we had to cut some. There were areas early on where there could have been territory issues, but I couldn't get over how well everyone worked together" (C. Spradling, personal communication). An important part of that teamwork involved frank discussions about concerns such as material covered in certain courses or other issues. "One of our biggest concerns was that we wondered how our students would fit in with the other departments," said Spradling. "For instance, what will the level of expectations be for our students from computer science when they take art classes?"

An area where the group could not reach agreement was the question of whether to include a course in creative photography in the core. Some faculty believed a course in writing for the professional media would be more valuable, while others believed it was important to emphasize the visual aesthetics covered in a photography course. But for the most part, the committee settled on the basics. Laura Widmer, Assistant Professor of Mass Communication, described the negotiating process in the following manner: "We came in here with one mission: to build a major that the industry would like and that our students would like and that we could offer" (Widmer, personal communication). Widmer also thought it was healthy that everyone in the group had input into which material should be covered in all three concentrations rather than delegating those decisions to faculty in each department.

Teamwork proved helpful for another reason as well: the seemingly never-ending volume of paperwork needed to move a new program through the lengthy approval process at the university level and beyond. In this case, there was even more paperwork than usual because two separate colleges were involved. Mass Communication and Computer Science/Information Systems are members of the Booth College of Professional and Business Studies, while Art is a member of the College of Arts and Sciences. "I always felt like we had administrative support for the new major," said Spradling, "but it wasn't just one set of papers we had to fill out. Instead, there were always two sets of papers we had to fill out." She added, "I don't think I would have done it by myself. The fact that we were a team was a big plus because you could divide up the work. You could take people whose strengths were in one area and farm out work for each person."

Near the end of the process, the ability to work together as a team became crucial in getting the new major approved at the state level. Another institution filed a challenge with the State Coordinating Board of Higher Education, claiming that it offered a similar program. The committee members were given two days to provide Northwest's leaders with ammunition for a response. They wrote a detailed course-by-course comparison of the two programs, which revealed that in key areas such as design, photography, electronic imaging, graphic design, broadcast operations, communication law, Web publishing, programming, marketing, television production, data structures, and networking, the program at the opposing college offered no course that was comparable to the proposed IDM major. The faculty also provided a list of "talking points" that summarized the industry need for the new major. Eventually the challenge was withdrawn.

In still another way, having two colleges in the mix was a bonus. Kim Spradling, Associate Professor and Chair of the Department of Art, observed, "Having three departments and two colleges involved helped get the major through the administrative hurdles because it was clear to everybody that the initiative was broad-based" (K. Spradling, personal communication). He added that without an interdisciplinary approach, a major such as this one would be impossible. "One department could not pull it off. If one department tried it would be duplicating or near duplicating courses. We were able to do this without adding any new courses." Even though there was little initial publicity about the new program, Spradling noted that half the high school students who visit his department ask about the new major. He expects the new major will be especially helpful for adding student credit hours in the Art Department because more students will take the foundation courses and because there are likely to be more classes offered in the summer. Getting students into those foundation courses can be a problem, however, unless enough sections are offered to handle the increased demand. To help launch the major, the university has authorized the three departments to hire a faculty member in instructional technology for a one-year term with the possibility of short-term renewal. Although no promises were made, Spradling expressed optimism that the numbers of students in the new major would eventually lead to additional administrative support in the form of equipment, software, faculty, and other resources.

Dr. Phillip Heeler, Professor and Chair of the Department of Computer Science/Information Systems, observes that the major grew out of a recognition that departmental boundaries are less all-encompassing than they used to be. For instance, Heeler said, "Although computer literacy is indispensable in the modern world, not everybody wants to major in computer science. There has got to be a middle road that includes the media or art and yet is based in computer skills" (Heeler, personal communication). Heeler elaborated by pointing out, "Freshmen come in and they really don't have a clue what they want to do. They might say they want to write computer games, but that's not a real viable option for a lot of people. They need to have additional, supplemental concepts and skills before they can even think about that. This cross-disciplinary major allows them to sample courses that will help them refine their goals and eventually find their niche."

As for the job market, Heeler sees opportunities in art and advertising as well as writing for the Web and more traditional computer science positions. When it comes to the entertainment industry, he observed that "the major motion pictures—*Pearl Harbor, Shrek*—are all so technology-driven. The audience is getting so sophisticated and our graduates need to be too." Job placement for graduates looks to be especially promising with high salaries and opportunities not only in the entertainment industry but also in corporate communications.

Heeler recognizes that the creation of a new major can present some administrative problems, especially when it comes to scheduling enough sections of introductory courses. "But if you look at the broader picture," he emphasizes, "if you have to offer four sections of this beginning level course, that's probably going to help later on because hopefully those students will stay around." Heeler adds that everyone will have to pay close attention to equipment requests, especially across colleges. Another potential problem—how to allocate student credit hours—appears to have been solved in this case by the creation of the three concentrations. Each department will claim as its majors those students enrolled in its specific concentration. Before taking on such an initiative, however, a department's leadership needs to carefully assess the likely impact on its SCH and on its number of full-time enrolled students.

ALIGNMENT WITH UNIVERSITY INITIATIVES

When designing a new program, one of the first steps in generating support from university administrators should be to demonstrate that the project is consistent with the institution's values and priorities as expressed in its vision and mission statement. In the case of the IDM major, the fit was as close to perfect as could have been hoped. Northwest has a reputation as "the electronic campus" that dates back to 1987, when the university instituted a program to install a computer in every student dormitory room and every faculty office. The IDM major seemed a natural extension and modernization of the university's commitment to utilizing new technology to enhance student learning and faculty achievement.

Northwest also is a leading practitioner of continuous quality improvement. The university was recognized in 1997 with a Missouri Quality Award in Education and was chosen for a site visit by the Malcolm Baldridge Quality Award Program. The university prides itself on fostering a "Culture of Quality," and that culture resulted in each of the three departments developing a professional advisory board made up of representatives from industry. These industry professionals enthusiastically supported the development of this new major to provide a new class of skilled employees for their industries.

The commitment to apply quality principles to improve life at the university was evident in the faculty's recognition of this emerging need and in the overall diligence of their research that justified the new major. The formal application makes these points in language familiar to university administrators when it states, "The proposed program in Interactive Digital Media supports the goals of Northwest Missouri State University's Vision and Mission Statements and Educational Key Quality Indicators by embedding interactive resources—human and technological—to provide seamless learning opportunities to diverse populations in a variety of settings. It furthers the University's position as a national leader in applying information technology to improve learning processes."

In order to back up its claim that the rationale for the new major was based on facts concerning media industry trends, the committee included a summary of its professional outreach efforts in the official application to the state to begin the new program. The document included a financial projection that included annual recurring expenses for faculty salaries and benefits, student employees, and upgrades to equipment and software. Although most everyone expected resources to become an issue someday, initial support of the proposal required no up-front funding because these projected expenses were expected to be offset by increases in tuition and fees from new students drawn to the major. As it turned out, the enrollment projections were conservative. One year after the program was instituted, the number of declared IDM majors totaled more than the amount anticipated at the four-year mark. The larger question to track in the near future would be how many of those students chose the university because of the new major and how many would have attended anyway.

ONGOING COLLABORATION

Successful interdisciplinary collaboration must be ongoing and not something that ends once the paperwork has been finalized. This point was expressed in different ways by all faculty involved in the initiative. Now that the new IDM major is in place and already attracting students, the greatest challenge will be for the faculty who designed it to continue to work closely together as well as to keep in close touch with the industry. "This major is very trendy, it's going to be very popular, and we have to stay current with our curriculum," said Widmer. "We have to keep bringing in professionals from the industry."

Effective advising of students will also require in-depth knowledge of course content from other departments. That might sound easy enough when a group has worked closely to design a program, but over time it becomes much more difficult as faculty revise their courses or leave for other opportunities. As Strauch put it, "I have to know what Craig Warner teaches in his Electronic Imaging class so that I can prepare students. That makes advising a lot more complicated than when you only have to know about your own department's courses." Faculty in each department will need to be kept informed about any relevant changes in course content and will also need to be proactive about recommending the inclusion of new ideas, trends, and technologies. The pursuit of quality quickly becomes an ongoing task with a continually receding horizon.

There is at least one silver lining, though, in the lengthy process of moving an initiative such as a new major through the myriad steps necessary to gain final approval. Warner cautioned that educators should not overreact to industry changes but instead need to be thoughtful before instituting a major academic response to a perceived need. "It's hard to be sure right away whether a change represents a major megatrend," he said. "You just have to listen to what people are saying and read a lot and most important of all give it some time. In this case, all the indicators seemed to be there even though for multimedia the Midwest is not the leading edge barometer of what's going on. Admittedly you would be better off going to San Francisco or Los Angeles or New York."

Still, Heeler sees cross-disciplinary programs as a trend that definitely will continue in higher education in the 21st century and new technology as the driving force behind that trend. "The thread of digital is tying us all together. There are digital ramifications and interfacing with so many disciplines now where there never was before. It has become a different world entirely."

CONCLUSION

Educators considering the development of an interdisciplinary program would be well advised to keep several priorities in mind. First, before forging ahead, ask the hard questions and be sure of the answers. Will the new program serve the educational needs of today's mass communication students better than the courses that any one department can offer by itself? Will the program serve a growing industry need, or is that perceived need merely a temporary phenomenon of the marketplace? Is it possible to generate sufficient support for the new initiative from the faculty involved and, of crucial importance, from the university administration? What are the projected start-up costs for the new program? Will it grow to be self-sufficient in time, or will a considerable, ongoing financial commitment be necessary in order to make the program successful?

Department chairs must also carefully weigh the impact of such a program on the numbers that administrators use when making decisions about number of faculty and allocation of resources. What will be the likely effect on student credit hours and full-time enrollment? In other words, will the initiative attract new majors from outside the departments involved as well as from outside the university? Or will current majors simply shift their areas of emphasis? Finally, and perhaps most important of all, can the faculty involved in the new program work together, both now and in the future? If not, the program is likely doomed to spin its wheels, leaving a legacy of interminable turf wars, hostility, and failure. Despite these considerable obstacles, the creation of the Interactive Digital Media major offers assurance that it can be well worth the effort for innovative educators to persist in creating new and powerful ways to prepare students for a world that will surely change.

REFERENCES

Heeler, P. (June 7, 2001). Personal communication.
Spradling, C. (May 31, 2001). Personal communication.
Spradling, K. (May 24, 2001). Personal communication.
Strauch, Jodell (May 22, 2001). Personal communication.
Warner, C. (June 6, 2001). Personal communication.
Widmer, L. (August 15, 2001). Personal communication.
Wissoker, K. (April 14, 2000). Negotiating a passage between disciplinary borders. *Chronicle of Higher Education*, B4–6.

Chapter 6

Bridging Academic and Professional Cultural Barriers in a Multimedia Reporting Class

Doug Sudhoff and Rick Musser

The assignment appeared simple in theory: take students from historically divergent academic tracks in a school of journalism, bring them together in one class, and lead those students beyond their chosen media to produce multimedia news products. In reality, that assignment met cultural barriers, both academic and professional. Those same barriers may challenge many universities now exploring media convergence in their own journalism classrooms. This chapter describes how faculty members in one school of journalism addressed those cultural barriers during the creation of a multimedia reporting class.

The faculty of the William Allen White School of Journalism and Mass Communications at the University of Kansas began preliminary discussion of curriculum revision in the early 1990s. The process took on new urgency with the arrival of Dean James K. Gentry in 1997. Under Gentry's guidance, a new curriculum that focused on gathering information and producing messages in a world of converging media emerged from faculty discussions. Over a period of three years, the faculty hammered out a plan that combined formerly separate undergraduate sequences (in newspaper; broadcast news, sales and management; magazine; business communications and advertising) into two tracks. One track was for students in news and information, the other for students with an interest in advertising and business communications. Both tracks were media neutral. The faculty's goal: make it easier for students to mold their own career paths in a rapidly changing media environment.

Prior to instituting the revised course of study, the school offered two experimental classes that reflected the goals of the new curriculum. The first, Research and Writing, introduced students to information gathering and writing skills they could apply to both informational (news) and persuasive (public relations and advertising) messages across a variety of media. The second, Multimedia Reporting, focused on bringing print and broadcast news students together.

Implementing Multimedia Reporting meant overcoming some long-established barriers within the school. Students in the broadcast and print sequences had attended very few classes together. Those few shared classes, such as media law and history of journalism, tended to be large lectures that offered the students little chance to interact across sequences. In addition, broadcast faculty resided in a different building, physically reducing interaction between print and broadcast students and faculty. Multimedia Reporting was born in an environment where the target students had shared few classes, few teachers, and few interactions. Those hurdles proved small compared to the academic and professional cultural barriers that conspired against the project.

REVIEW OF LITERATURE

The Multimedia Reporting project began with the understanding that a changing media landscape demands different training for future journalists. Thirty-seven percent of Americans between the ages of 30 and 49 use the Internet at least once a week for news, while 18% of that group goes online daily for news ("Internet Sapping Broadcast News Audience," 2000). According to the Pew Center survey, an even greater number of Americans under the age of 30 use the Internet as a source of news. The Center's research also shows that the growing use of online sources for news corresponds with a decline in television news viewing and little or no growth in newspaper readership.

The research shows that many news consumers have abandoned traditional media loyalty. A growing number of people seeking news use a variety of media to find the information they want. (Hammond, Petersen, and Thomsen, 2000). Taken to its extreme, the growing ability of individuals to gather information from the Internet means everyone with access to an online workstation could become his or her own news editor (Oreskes, 1999).

This changing environment creates challenges for both journalists and their news organizations. First, journalists must reassert their basic value to society. They must show that they provide a special service—fair, balanced, and useful information—that goes beyond what members of the general public can achieve for themselves (Oreskes, 1999; Winch, 1997). At the same time, they must develop a set of skills that allows them to work across a spectrum of media, many of which are converging on the Internet.

A growing number of media companies now seek cooperation among their print, broadcast, and online outlets, reinforcing the need for journalists to acquire a broader set of skills. Companies such as the Tribune Company are leading the way in sharing information between their print and broadcast holdings (Hammond, Petersen, and Thomsen, 2000). Gentry (1999) writes that the Tribune Company's *Orlando Sentinel* has made "a total

commitment to convergence" with newspaper, television, and web operations sharing information and cross-promoting one another. While, admittedly, few fully integrated multimedia newsrooms currently exist, a growing number of news operations, as well as other information industry employers, have begun to seek out employees with skills that allow them to operate across a variety of media (Hammond, Petersen, and Thomsen, 2000; Griffin and Pasadeos, 1998). Andersen (2000) notes that converged media means presentations that offer text, photographs, audio, and video simultaneously. With this in mind, reporters for the *Chicago Tribune*'s Internet operation must carry "a pen, a notebook, a tape recorder for audio clips, a digital camera for single snapshots and sometimes a video camera for video clips" (Harper, 2000). At the *Orlando Sentinel,* the growing cooperation between the newspaper operation and the television operation has led to "the expectation that (every newspaper reporter) must participate" in providing content for both the paper and the news broadcasts (Gentry, 1999).

But, while a growing number of news organizations seek journalists with multimedia skills, are institutions of higher learning prepared to help undergraduates fulfill those needs? Pasadeos (2000) notes that many universities integrate their mass communications curriculum to the extent of offering some common courses such as ethics, research, and law. There are fewer instances of integrated skills courses. Over the past five years, some universities—such as Brigham Young University, Marshall University, and Virginia Commonwealth University—began to develop integrated, multimedia newsrooms for students (Hammond, Petersen, and Thomsen, 2000). However, it appears few universities have attempted to bring print journalism and broadcast journalism students together early in their undergraduate careers to begin developing multimedia reporting skills. Perhaps that is because any attempt to do so faces a number of academic cultural barriers.

First, many mass communication educators contend that requiring students to learn skills beyond those in a chosen, specialized arena—for example, broadcast journalism or copy editing—makes it more difficult for the students to become fluent in a single discipline (Kock, Kang, and Allen, 1999; Griffin and Pasadeos, 1998). Proponents for divergence contend that developing multimedia journalists will result in graduates entering the job market with a reduced set of job skills.

Second, the academy may make it more difficult for faculty members to break down barriers within their own school. As Griffin and Pasadeos (1998) note, universities reward instructors who devote their careers to in-depth study of a particular discipline. In particular, many tenure-track faculty find that focusing their work on a specific area of mass communication provides the safest route on the long journey to tenure. This focus on specialization makes research, teaching, and service in an area different than an instructor's original medium very difficult.

Finally, Pasadeos (2000) says this emphasis on specialization means that mass communication faculty rarely read or conduct research outside their particular areas. This insular environment heightens the academic barriers and retards any movement toward creating a class that teaches both print and broadcast reporting skills.

These cultural barriers in the academy foreshadow the professional cultural barriers that exist once students graduate. A reporter at the *Orlando Sentinel* noted that she felt very skeptical when the paper began its move toward a converged news environment because "print and television always seem to have an antagonistic relationship" (Gentry, 1999). Hollifield, Kosicki, and Becker (2001) say that while journalists in different media hold some shared values—such as fairness—they also maintain values inherent to their particular medium. Thus, journalists at a newspaper have historically operated under a different set of expectations compared to their counterparts in a television newsroom.

Most newspapers adhere to a 24-hour news cycle. Although there may be various editions, a newspaper comes out essentially once each day. Broadcast news outlets—and the growing number of online news providers—demand frequent, and in some cases constant, updates of information for the audience (Hammond, Petersen, and Thomsen, 2000; Harper, 2000; Oreskes, 1999). This aspect of electronic journalism means reporters, editors, and producers have far less time to gather information and produce coherent, thoughtful narratives.

Whitaker, Ramsey, and Smith (2000) say that television news operations tend to use reporters as "generalists" as opposed to specialists. Newspaper reporters focus on one "beat" far more than TV reporters. A newspaper journalist may cover the police or the courts or city hall. A television reporter—especially in small-to-medium size markets—will cover all those venues as well as others. This, in part, reflects the fact that TV news operations generally employ fewer journalists than do newspapers. That difference is accentuated as market size grows.

Television journalists also tend to face greater daily pressure to provide content than do their print counterparts (Hammond, Petersen, and Thomsen, 2000). If a newspaper story falls through, editors may expand or contract the local news budget to reflect the number of narratives available. Television newsrooms, with a day-long parade of half-hour blocks that must be filled, regardless of the number of stories and reporters available, operate at a more harried pace. TV reporters adopt an "I need to get *something* for the newscast" attitude, a responsibility accentuated by the smaller staff size of TV newsrooms.

Thus, undergraduate preparation for the journalism profession occurs in a culture where newspaper reporters receive more time to research and produce their narratives than do television reporters. Even as students, print journalists tend to specialize in a particular kind of coverage more than do television reporters. Although they may argue the point, print reporters tend to face less pressure to provide content on a daily basis than television reporters. These professional cultural differences, and the academic cultural barriers discussed earlier, all were present at the creation of the Multimedia Reporting class at the University of Kansas.

IDENTIFYING ACADEMIC AND PROFESSIONAL CULTURAL BARRIERS

The initial Multimedia Reporting class convened in the spring of 2000 with the goal of introducing students in the print journalism and the broadcast journalism sequences to an intermedia approach to reporting. While long-term plans called for eventually introducing an Internet component to the class, the first group of students in Multimedia Reporting focused on producing narratives for print and television. To accomplish this goal, each student produced a series of assignments, some for the print medium, others for television. The final project called for each student to produce a story for both print and television on the same topic. The instructors outlined in the syllabus standards that reflected A, B, C, and D work. A-grade standards met publication guidelines for the student newspaper, the *University Daily Kansan,* and broadcast guidelines the school's student-produced nightly newscast on Channel 14.

A two-person teaching team—one instructor from the print sequence and one from the broadcast sequence—met regularly during the months prior to launching the class. During those meetings the instructors established that, on any given story, (1) the TV narrative would include fewer details than the print version, (2) the TV narrative would take into account the video available to the reporters, and (3) the writer of the TV narrative would adopt a simpler text style than the writer of the print narrative due to the "here and gone" nature of any report airing on television.

As course planning progressed, the team identified a series of cultural "myths." Overcoming these myths played an important role in breaking down the barriers that, at the beginning of the semester, existed between students, the two sequences, and even the instructors themselves.

According to the first myth, newspaper reporters rely on the inverted pyramid narrative, while television reporters rely on a more conversational writing style. While the inverted pyramid enjoys a long history in print journalism, the team developed a better narrative model based on active voice, short sentences, and compelling leads, a model that the two media easily shared.

The team addressed the second myth, that newspaper journalists write in the past tense while television journalists write in the present tense, by stressing that overreliance on such rules produced bad writing in both media. The instructors demanded "forward spin" on stories, telling students writing for either medium to look beyond what had already happened and to "advance" the story as part of their narrative. Students writing television narratives were warned to avoid what the instructors identified as "false presentism," the tendency to force past events into the present tense. Students writing print features were encouraged to experiment with present tense attribution.

The next myth insists that print journalism requires more thought and effort than broadcast journalism. The teaching team acknowledged that print narratives tend to run longer and include more sources than do television narratives. As such, newspaper stories tend to be more complex than typical television reports. However, both teachers stressed that good TV journalism requires that the reporter blend narration, video, and audio. Thus, television reporters—already facing a compressed time line for producing their narratives—must create stories that incorporate far more elements and far more production steps than the written word.

That left the cultural, and not entirely mythical, perception that print represents a generally more respectable form of journalism than does television. To deal with this barrier, both instructors acknowledged the criticisms leveled at print (that the medium can be staid and lags behind broadcast and Internet sources in timely information) and television (that the medium can be shallow and sensational). In the end, the instructors simply ignored these ideological flashpoints. Instead, classroom efforts focused on mutual respect and producing solid, useful reporting in both media while working within the constraints imposed by the industry.

TEACHING METHODS ADOPTED FOR MULTIMEDIA REPORTING

Successfully breaking down professional cultural barriers in multimedia reporting also meant addressing the academic barriers between sequences in the school of journalism. A detailed examination of the process used to accomplish this at the University of Kansas goes beyond the scope of this article. However, a brief discussion of three key concerns is useful.

The first barrier was the widely held concern that developing multimedia skills would rob undergraduates of highly developed skills for a particular medium. That problem was primarily addressed in the sequencing of courses in the new curriculum. The school's revised curriculum requires that, once news students complete a core of common study—including multimedia research and writing, ethics, First Amendment issues, and multimedia reporting and editing—they must take medium-specific courses. Those professional skills courses still require students to demonstrate skills necessary for a specific medium.

To overcome entrenched specialization, the William Allen White School sought to encourage and reward cross-media teaching. Part of the faculty-devised curriculum revision called for all faculty members to, at various times, teach classes in the multimedia core. In essence, the faculty built in requirements to move outside the box of their own specific interests.

This speaks to the final concern, that faculty often lack experience working outside their area of specialization. In this case, faculty members adopted a system of internal mentoring, where instructors from one sequence helped those from another sequence grade assignments and develop the basic skills needed to work with undergraduates in a multimedia environment.

In Multimedia Reporting, the teaching team determined that students must see that, regardless of medium, good writing and good reporting lie at the heart of any good news narrative. To that end, the instructors adopted several strategies aimed at reinforcing common points of good storytelling for both print and television. First, the teachers adopted the *Wall Street Journal* story model for both print and television narratives. Rich (1994) describes the *Wall Street Journal* model as storytelling that moves from specific to general, beginning with an anecdotal lead that generally focuses on a person or event that exemplifies the main issue. The narrative then moves to a fuller discussion of the issue and concludes by coming back to the person or event that provided the anecdotal lead. This model—while not appropriate for every news narrative—provided a format that both print and broadcast students could use. It also gave the students a common point of reference regardless of the medium in which they were working.

The teaching team also identified and stressed writing rules common to both media. To this end, the instructors constantly encouraged the students to use the subject-verb-object sentence construction, to use active voice verbs, and to make sure they chose words with specific, concrete meaning. In a nod to the needs of television news, the team encouraged students to avoid compound sentences, to avoid stating ideas in the negative, and to keep sentences short and easily understood. These rules focused on the goal of clearly communicating critical information—regardless of the medium of transmission.

Students in Multimedia Reporting experienced the demands of both media. Students from the broadcast sequence learned to write longer narratives that wove in more sources than the typical TV report. Print students learned to write shorter narratives that the audience would hear and see, rather than read. All students learned to develop fundamental skills in videography and video editing.

To achieve these goals, the teaching team spent the first quarter of the semester addressing fundamental skills. The students practiced basic videography (framing, lighting, and sequencing) and video editing skills (visual logic, pacing, and transitions) for both b-roll and interviews. While new for the print students, this step provided a useful review for the broadcast students. The students also wrote frequently. The assignments emphasized the writing rules discussed earlier and introduced the *Wall Street Journal* model.

The students generally worked in pairs. Where possible, the instructors teamed a print student with a broadcast student. This allowed the students to mentor one another. Print students often helped broadcast students with the style or sourcing requirements of reporting for newspapers. Broadcast students helped their counterparts from the print sequence with the intricacies of producing video for television.

The class spent the remainder of the semester working on various news narrative genres common to all media: an animal feature, the business report, science report, and sports report. These "evergreen" narratives allowed the students more time for research, videotaping, editing, and writing than attempting to cover breaking news. The teachers used a two-week block for each genre. Each block included one lecture with examples to introduce the particular type of narrative, brainstorming sessions for each student team to develop story ideas, time for each student team to research a particular story idea, and a class session in which teams shared story ideas, with other students in the class critiquing the proposals.

Once assigned a story idea, each student team began information gathering, shooting and editing video, and writing both print and television versions of the report. The print version generally required a certain number of column inches, while the TV version generally took the form of a one-minute voice-over/sound bite (VOB). At the end of the two-week block, each team shared its print and television versions with the rest of the class, which then critiqued the work. The teams shared grades for both versions of the story.

The students in each team alternated working in print and broadcast. Thus, a student who produced a print narrative for one genre would produce a television narrative for the next genre. Students received bonus points for stories that appeared in the *University Daily Kansan* and on Channel 14 News. For the final project, each student produced both a print and television version of an enterprise story he or she developed. Final projects earned grades on an individual rather than team basis. The students began working on the final project at mid-semester, giving them about seven weeks to complete their work.

DISCUSSION

The most fundamental result of the multimedia reporting experiment was proof that teaching such a course was, indeed, possible. In the semesters following the pilot course, the school adopted the model developed by the teaching team as the basis for multiple sections of the required Journalism 415 Multimedia Reporting. While modifications certainly did, and will, take place (e.g., the final project became an online, multimedia product; the sports story became a sports profile; all students were assigned a spot, police news shift for the student radio station), the basic structure and content of the course remains unchanged.

What made the course viable was willingness on the part of both teachers to find common ground on reporting and writing goals. Neither teacher began with an everything-I-teach-must-stay-in-the-course attitude. Both continually asked: What is it that students really need to know before they go on to the next, specialized media course? Both instructors also were willing to acknowledge that such basic skills would need to be refined in later, media-specific courses.

In the end, the teachers found more common ground than perhaps either had anticipated. Learning news skills and finding new connections in reporting and writing news also energized both teachers.

While teaming any students inevitably produces some personnel conflicts, a surprising number of students with divergent media goals did mentor and help one another. And, while some high-GPA print students with stunted visual skills grumbled, the highest praise in student feedback often was "This wasn't as bad as I thought it would be." Anecdotally, the class produced a number of successes. Students from the class continued to have their work published in the student newspaper with a frequency not noticeably different from students in the old, print-specific, reporting class. Student work from the class also aired on the campus television evening news. Students went on to function normally in student media. Teachers in the subsequent, advanced student media courses cautiously observed that TV students who had taken Multimedia Reporting seemed better prepared in their reporting skills and that newspaper reporters educated under the old print and new multimedia style appeared indistinguishable. There was also some anecdotal evidence that good print reporters benefited from writing lessons learned in the TV style, producing leads containing more visual elements and immediacy.

The course was not without its detractors. The constant switching between print and broadcast copy produces students less mindful of "correct" style, tense, and attribution. That has not been popular in a program that prides itself on producing copy editors for newspapers. But, while student editors (perhaps acting as proxies for some faculty) have penned dire warnings about the oncoming devaluation of their journalism degrees, students at KU won more Dow Jones editing internships in the summer of 2001 than at any time in the school's history.

What the faculty has yet to work out is the impact of multimedia reporting on the way students are taught in the advanced student media classes. Under the old curriculum, the equivalent of multimedia reporting was the last required course a news student had to take. Under the new curriculum, all students in the news track must take two student media courses, which formerly were electives.

What is clear is that students who learn cross-media reporting and editing skills are capable of going directly to work on the student newspaper and TV station. What is not clear is how the new curriculum will affect a system that has relied on students to run their own media. The fact that every news student now must work as a reporter, producer, or copy editor has yet to percolate through a student media environment still gripped by the same cultural barriers that Multimedia Reporting seeks to overcome.

But the Multimedia Reporting pilot course proved that such barriers are not insurmountable. Among faculty, planning and team teaching the multimedia courses continue to open the way for more cross-media communication among news faculty with broadcast and print backgrounds. Some changes are dramatic: The print half of the pilot course teaching team, who had 25 years of experience working and teaching in newspapers and magazines, spent the summer of 2001 working in a television newsroom in Chicago. Some changes are more subtle: Students who manage the campus TV station and newspaper will soon all share the same exposure to faculty and each other. Both types of changes pave the way for the next challenge—creating student newsrooms that function in a truly multimedia mode.

REFERENCES

Andersen, Neil. "Media Education and the New and Converging Technologies." *Media Launchpad.* http://www.media-awareness.ca/eng/med/class/andersen.htp

Gentry, James K. *Newspaper of the Future: Integrating Print, Television, and Web.* Reston, VA: American Society of Newspaper Editors, Change Committee, 1999.

Griffin, W. Glenn, and Yorgo Pasadeos. "The Impact of IMC on Advertising and Public Relations Education." *Journalism and Mass Communication Educator* 53 (Summer 1998): 4-18.

Hammond, Scott C., Daniel Petersen, and Steven Thomsen. "Print, Broadcast and Online Convergence in the Newsroom." *Journalism and Mass Communication Educator* 55 (Summer 2000): 16–25.

Harper, Christopher. "Doing It All." *American Journalism Review* 18 (December 1996): 24–30.

Hollifield, C. Ann, Gerald M. Kosicki, and Lee B. Becker. "Organizational vs. Professional Culture in the Newsroom: Television News Directors' and Newspaper Editors' Hiring Decisions." *Journal of Broadcasting and Electronic Media* 45 (Winter 2001): 92–117.

Internet Sapping Broadcast News Audience. *The Pew Research Center for the People and the Press.* Retrieved July 10, 2001, archived at www.people-press.org/report/display.php3?reportid=36

Kock, Erin, Jong G. Kang, and David S. Allen. "Broadcast Education Curricula in 2-Year and 4-year Colleges." *Journalism and Mass Communication Educator* 54 (Spring 1999): 4–15.

Oreskes, Michael. "Navigating a Minefield." *American Journalism Review* 21 (November 1999) 20–25.

Pasadeos, Yorgo. "Conflicting Attitudes Toward An Integrated Curriculum." *Journalism and Mass Communication Educator* 55 (Spring 2000): 73–78.

Rich, Carole. *Writing and Reporting the News: A Coaching Method.* Belmont, CA: Wadsworth Publishing, 1994.

Whitaker, W. Richard, Janet E. Ramsey, and Ronald D. Smith. *Media Writing: Print, Broadcast and Public Relations*. New York: Longman, 2000.

Winch, Samuel P. *Mapping the Cultural Space of Journalism.* Wesport, CT: Praeger, 1997.

CHAPTER 7

The Changing of Teaching-Learning Relationships in the Classroom: Exploring the Transition to Teaching Digital Technologies

Joyce Chen

INTRODUCTION

Many students enrolled in the electronic media major are no longer purely attracted to broadcasting. Instead, they are fascinated by the versatility of computerized media and eager to learn the knowledge and skills needed to gain a competitive edge in the ever-changing electronic media industries (Chen & Bankston, 1999). Accordingly, the question of how best to teach digital technologies has become an inevitable challenge to electronic media educators.

Instructors who teach electronic production courses experience the effects of the transition from analog to digital technology on a daily basis. Especially, due to rapid upgrades of computer hardware and software (Van Tassel, 2001), digital technology constantly challenges instructors to update their knowledge. At issue is more than whether an instructor needs to change his or her syllabus from teaching analog to teaching digital techniques. There are also pedagogical, cultural, and aesthetic questions. The discussion has been under way for some time through convention panels and online listservs, but to better understand the current state of the academy it is necessary to examine it further. This chapter addresses the issue of how the teaching-learning relationship changes as electronic media instructors adapt to the digital age. Related issues include the impact of the digital age on changes in equipment/format selection, learning skills, financial resources, and administrative and technical support (Barr & Tagg, 1995; Green & Gilbert, 1995; Fink, 1998).

LITERATURE REVIEW

With the wide application of media technologies to every field in society, there is an increasing demand for advanced electronic media education (Chen & Bankton, 1999; Regan, 1995). Brunner and Tally (1999) assert, "Media technologies greatly expand the range of teaching, learning, and communication modalities available to teachers and students, but at bottom their only value is in helping us to do other things well" (p. 22). They believe media technologies make a democratic, information-age school possible where students' interactions with teachers and peers, as well as their self-initiated learning skills, are greatly promoted. Green and Gilbert (1995) discuss intensively the relationships between new technologies and productivity, content, and communication in higher education. Their study shows that new technology has made "a qualitative difference in the way we teach, the materials we teach with, the structure of the college curriculum, the learning experience for students, and how we exchange information—both with colleagues and as faculty interacting with students" (Green & Gilbert, 1995, p. 18). Connick and Russo (1995) address the relationship between technology and inevitable educational transition. They emphasize three major educational issues: access, quality, and productivity. The application of technology has altered traditional methods of teaching and learning. In order to promote educational access, quality, and productivity in the information age, support from the institution's administrative leadership is the key. They claim, "Leadership can funnel resources, support, and tools into the hands of those who will effect real change—the institution's faculty" (Connick & Russo, 1995, p. 18).

Accordingly, the school system needs to change too, especially the teachers' role and the creation of the new role of media and technology specialists. Connick and Russo argue that teachers should "leave fact-finding to the computer, spending their time doing what they were meant to do as content experts: arousing curiosity, asking the right questions at the right time, and stimulating debate and serious discussion around engaging topics" (p. 33). Therefore, "teachers cannot be expected to keep up with the latest software upgrades, understand the newest vogue in online interactions and its implications for classrooms. . . . On-demand school-based technology support is therefore crucial. But the most common organization of technology support in schools—"the media specialist" or computer coordinator who presides over the computer lab and has very little to do with teachers or curriculum—is a key barrier to successful technology integration" (p. 38). Regarding technology and learning, Bruffee (1999) argues that "to devise software that can take these initial steps toward genuinely interactive educational software, designers would strive first to weaken the bond that glues students to the keyboard and mouse by intensifying the excitement, interest, and understanding that students derive from conversation about what they are seeing or have seen on the monitor" (p. 126).

However, Green and Gilbert (1995) do not find evidence that there are institutional gains in productivity. Institutional and instructional productivity means to maintain and enhance student learning while significantly reducing instructional cost, i.e., effective cost control. As a matter of fact, the improvements and increased values in higher education offered by new technology are usually in conjunction with substantially increased investment and support costs, including hardware, software, faculty time, support services, and so forth (Green & Gilbert, 1995, p. 18). More important, new technology has transformed higher education in achieving new levels of efficiency and effectiveness, in terms of teaching, learning, and the curriculum. Kozma and Johnston (1991) identify seven types of transformation: (a) from reception to engagement; (b) from the classroom to the real world; (c) from text to multiple representations; (d) from coverage to mastery; (e) from isolation to interconnection; (f) from products to processes; and (g) from mechanics to understanding in the laboratory (pp. 16–18).

Steele (1995) believes that confidence, cost, and convenience are three important factors that influence instructors' attitudes toward new technologies (p. 23). He finds that "most who teach would prefer to do the best job possible. Often, they retreat to old habits and stop trying because they cannot count upon convenient, dependable, and easy-to-use support systems" (p. 28).

Fink's winning paper (1998) at the BEA convention in 1998 investigates the perceptions of broadcasters and educators about the changing environment that digital technology brings to media production education. His qualitative study finds four areas that digital technology impacts: (a) the production process, (b) curriculum, (c) class interaction, and (d) budgeting. Digital technology not only greatly improves the quality, aesthetics, and scale of production, but also revolutionizes electronic media production. It does so by expanding broadcasting to include multimedia production and distribution, by networking multimedia outlets to global geographical locations, and by replacing the linear process of production and distribution with the nonlinear process and with video-on-demand reception. Accordingly, the curriculum in electronic media education must incorporate digital technology and these new aesthetic concepts of production. Further, the preparations of teaching and student-instructor interactions in and out of the classroom need to be investigated and reconsidered. His respondents agree: "Digital technology opens many creative possibilities in curriculum design. However, instructors often find they need to spend more time than they would like on skills training to the detriment of teaching the theoretic concepts of [digital] production" (Fink, 1998).

Moreover, digital technology promotes the process that the traditional teaching-centered education has shifted to learning-centered, i.e., the teaching paradigm shift from lecturing to interactive learning (Barr & Tagg, 1995; Fink, 1998). Barr and Tagg (1995) advocate for changing undergraduate education from the Instruction Paradigm to the Learning Paradigm. They define that the mission of college education in the Learning Paradigm is to produce learning rather than to provide instruction. It shifts a college's responsibility from quality instruction (lecturing and talking) to the creation of a powerful environment for student learning. Students as the coproducers of learning must take responsibility for their own learning. The Learning Paradigm is student-centered, honoring whatever approaches serve best to prompt learning of particular knowledge by particular students. It promotes cooperative, collaborative, and supportive learning. It pays more attention to the quantity and quality of student-success outcomes rather than quantity and quality of entering students or instructional resources. In the Learning Paradigm, faculty members are primarily designers of learning methods and environments rather than lecturers who deliver "chunks" and "bits" of knowledge to students. Teachers need to work together with students to develop competencies and talents rather than to classify and sort students. At the same time, students need to learn through individual experiences and by developing analytical thinking (Barr & Tagg, 1995).

Fink (1998) asserts that the electronic media instructor's role is no longer a disseminator of knowledge and skills but an interactive facilitator who initiates, guides, and oversees the relationship between new technology and education, the interactions of students with instructors, and of students with students. Therefore, instructors need to be trained in digital technology, including audio, video, and multimedia production, and in the instructional design and new paradigm of teaching. On the other hand, students need to understand that "learning must be a process of discovery through cooperative efforts with others," and be responsible to continuously extend their digital knowledge and skills by engaging more in self-education under their instructor's supervision (Fink, 1998). Competency in computer skills is essential for students to be successful in getting an entry-level position or promotion in the workplace (Fink, 1998; Lind & Danowski, 1999). Although currently broadcasters like to train students in high-end software applications on the job, in the near future electronic media employers will tend "to hire graduates who are both conceptually literate and high-end application skilled because [many] universities and trade schools are turning out more and more graduates who have those skills" (Fink, 1998).

Gordon Moore, the cofounder of Intel, states that the capacity of microprocessor chips would double every 18 months (Van Tassel, 2001). In order to stay current, budgeting 30% of the original purchase price each year is necessary to keep up with upgrades, new software, maintenance, and training (Fink, 1998, p.13). However, as a matter of fact, "most universities amortize equipment purchases over three to eight years" which is too long for digital technology (Fink, 1998, p.13). Thus, the crucial challenge to electronic media education is "how to afford enough workstations for the students to have a meaningful experience, while still emphasizing the conceptual basics of production, and fit the department's meager budget" (Fink, 1998, p.13). Further, support staff and faculty usually need to be trained on their own time and cost in order to stay one step ahead of the students with each new and upgraded machine and application (Fink, 1998, p.13).

These studies provide insightful references for the present research and help to conceptualize the changing teaching-learning paradigm.

RESEARCH QUESTIONS

In order to explore the impact of digital production technology on teaching-learning relationship in the classroom, two research questions were posed:

> RQ1: What are students' perceptions of digital technologies?
>
> RQ2: What are the pedagogical differences between teaching digital production and teaching analog production?

RESEARCH METHODS

With a combination of quantitative and qualitative approaches, this study investigates the experiences and problems of electronic media instructors and students in adopting digital production technology. It also provides an overview of the transition process among BEA schools. Thus, one key to the success of this study is the participation of instructors and students among the BEA schools in the discussion.

INSTRUMENT

There were two versions of online web survey questionnaires designed by using Microsoft FrontPage. One was for instructors and the other for students. Both questionnaires consist of questions regarding digital production systems used in schools, and respondents' perceptions of digital systems. The one for instructors also includes questions about instructors' experiences in teaching digital technologies and their perceptions of students' attitudes toward learning digital technology. The one for students also asked questions about students' attitude toward digital technologies and their perceptions of the impact of digital technology on electronic media instructors' teaching.

The questionnaire was designed based on the following considerations: How long has electronic media education been using digital technologies? What kinds of digital audio/video equipment/facilities were used in electronic media production classes? Compared to analog technologies/facilities, what are the advantages and disadvantages of digital electronic media production, in terms of access, quality, and productivity? What problems (regarding confidence, cost, and convenience) have instructors encountered in adopting digital production technologies? How did they attempt to solve those problems? How did instructors learn digital production knowledge and skills? What needs to be done to best teach digital production? What do instructors expect for the future development of digital production technologies? What are students' attitudes toward digital technologies?

In addition, the survey included comment sections to encourage faculty members and students to express their individual experiences and opinions on using digital production technologies for further interpretation of the survey results.

RESULTS

There are 72 faculty respondents from 44 institutions and 205 student respondents from three institutions. Among 205 students, there are 103 females (50.2%) and 102 males (49.8%), and 128 (62.4%) students are electronic media majors. There are 9 freshmen (4.4%), 33 sophomores (16.1%), 58 juniors (28.3%), and 96 (46.8%) seniors. About 80 percent of the student respondents began using digital technologies in 1994 or later, and about 50% of the students started learning digital technologies by taking courses.

Among the 72 faculty respondents, 47% are teaching in institutions that have installed digital equipment during the past 1 to 5 years, 33% during the past 6 to 10 years, 11% during the past 11 to 20 years. Nine percent have no digital system yet. About 58% of the faculty have used digital technologies for 1 to 5 years, about 17% for 6 to 10 years, and 20% for 10 to 20 years. Five percent had no experience using digital technologies.

RQ1: What are students' perceptions of digital technologies?

In Table 7.1, the mean scores show that students perceive more advantages to digital technology than faculty perceive, especially such indicators as "easier to learn" ($t = 4.58$, $p = .000$), "easier to operate" ($t = 4.95$, $p = .000$), "easier to set up" ($t = 5.11$, $p = .000$), and "demonstrating operation easily" ($t = 2.59$, $p = .010$). However, except for "better for dubbing" ($t = -3.86$, $p = .000$) faculty members experience the need for more extra

Table 7.1
Comparison between Faculty and Students (*t*-test)

Perceptions of Digital Technology	Mean (n) Faculty	Mean (n) Students	2-tail *t*	2-tail *p*
More productive	2.06 (70)	1.88 (193)	1.36	175
More accessible	2.47 (68)	2.41 (194)	.42	.678
Easier to learn	2.99 (69)	2.32 (193)	4.58	*.000*
Easier to operate	2.85 (68)	2.14 (193)	4.95	*.000*
Higher production quality	1.80 (70)	1.85 (193)	– .39	.695
More time consuming	2.55 (69)	3.20 (193)	– 4.06	*.000*
Better for dubbing	1.51 (70)	2.03 (193)	– 3.86	*.000*
More difficult to teach	2.91 (69)	3.04 (193)	– .79	.429
Easier for students to learn	2.69 (70)	2.46 (192)	1.63	.105
More difficult to handle technical problems	2.28 (71)	2.77 (193)	– 3.60	*.000*
Easier to set up	3.11 (71)	2.41 (193)	5.11	*.000*
More expensive to upgrade the system	2.62 (68)	2.93 (192)	– 2.05	*.041*
Facilitating students' creativity	2.06 (69)	1.94 (203)	.84	.404
Stimulating students' curiosity	2.22 (69)	2.01 (203)	1.47	.142
Requiring more individualized instruction	2.21 (68)	2.73 (203)	– 2.93	*.004*
Requiring more assistance for students in the lab	2.23 (69)	2.53 (202)	– 1.98	*.048*
Demonstrating operation easily	2.88 (67)	2.48 (203)	2.59	*.010*

Note: 1 = Strongly Agree, 2 = Agree, 3 = Neutral, 4 = Disagree, 5 = Strongly Disagree

effort in teaching digital technology than students believe necessary, such as "more time consuming" (t = -4.06, p = .000), "more difficult to handle technical problems" (t = -3.60, p = .000), "more expensive to up-grade the system" (t = -2.05, p = .041), "requiring more individualized instruction" (t = -2.93, p = .004), and "requiring more assistance for students in the lab" (t = -1.98, p = .048).

In Table 7.2, the responses indicate that faculty encounter problems with implementing digital technology that are different from student experiences. Faculty members perceive the "lack of formal training" to be more serious than students experience it (t = -4.75, p = .000). Students are more sensitive than faculty to such problems as "lack of support from administrators" (t = 2.66, p = .008), and "system's functioning not well" (t = 2.35, p = .020). Although the mean scores indicate that both faculty and students disagree with the statements such as "not feeling comfortable to use digital technology" and the "quality not good enough," faculty confidence in using digital technologies is significantly higher than students',

Table 7.2
Perceptions of Problems: Comparison between Faculty and Students (*t*-test)

	Mean (n)		**2-tail**	
Perceptions of Problems	**Faculty**	**Students**	*t*	*p*
Lack of funding	2.39 (72)	2.43 (192)	– .28	.779
Lack of support from administrators	3.25 (72)	2.90 (191)	2.66	*.008*
Lack of knowledge about digital systems	3.46 (72)	3.17 (193)	1.93	.054
Lack of technical support	2.75 (72)	2.82 (191)	– .45	.650
Lack of formal training	2.35 (72)	3.04 (191)	– 4.75	*.000*
System's functioning not well	3.33 (72)	2.99 (189)	2.35	*.020*
Price too high	2.94 (72)	2.77 (189)	1.29	.199
Not feeling comfortable to use digital technology	3.74 (72)	3.22 (190)	3.77	*.000*
Quality not good enough	3.65 (72)	3.12 (189)	3.52	*.001*

Note: 1 = Strongly Agree, 2 = Agree, 3 = Neutral, 4 = Disagree, 5 = Strongly Disagree

Table 7.3
Perceptions of Student Attitudes Toward Digital Technology (Percentages of Strongly Agree and Agree)

Student Attitudes Toward Digital Technology	Faculty (%)	Students (%)
Eager to learn digital technology	71.2	89.4
Curious to explore digital technology	73.5	89.4
Afraid of operating digital technology	14.7	15.6
Frustrated when encountering problems	41.8	43.3
Optimistic toward digital technology	71.6	78
Not interested in digital technology	3	7.1
Confused by digital technology	23.5	25.5

i.e., "not feeling comfortable to use digital technology" ($t = 3.77$, $p = .000$) and "quality not good enough" ($t = 3.52$, $p = .001$).

Table 7.3 demonstrates the differences (10% or higher in agreement) about students' attitude toward learning digital technology between faculty's perceptions and students' self-reports. Especially for those positive attitudes, such as "eager to learn" and "curious to explore," students report their attitudes toward learning digital technology more enthusiastically (89.4% for both attitudes) than those perceived by their instructors (71.2% for "eager to learn," and 73.5 for "curious to explore").

There are significant differences between electronic media majors and other communication majors, regarding the questions "better for dubbing," "lack of support from the administrators," and "confused by digital technologies." Electronic media students appreciate more about dubbing qualities ($t = -2.77$, $p = .006$), perceive less "the lack of support from the administrators" ($t = 2.44$, $p = .016$), and are less "confused by digital technologies" ($t = 2.74$, $p = .007$) than students not majoring in electronic media. There are gender differences. Male students perceive more "accessible to the systems" ($t = -2.69$, $p = .006$), as well as "easier to operate" ($t = -2.05$, $p = .042$), "better for dubbing" ($t = -2.59$, $p = .010$), less funding problems ($t = 2.56$, $p = .011$), less "need for assistance in the lab" ($t = 2.82$, $p = .005$), less "easier to demonstrate" ($t = 2.07$, $p = .039$), less "afraid of operating digital technology" ($t = 2.35$, $p = .020$), and less "confused by digital technology" ($t = 2.71$, $p = .008$).

One-way statistical results show that there are significant differences among students with various years in college education, regarding accessibility of digital technology (n = 175, $F = 4.4$ $p = .014$) and the need for individualized instruction of digital technology (n = 185, $F = 3.88$ $p = .022$). Further t-tests show that junior students experience digital technology more positively than students in other grades.

RQ2: What are the pedagogical differences between teaching digital production and teaching analog production?

Factor analysis is employed to examine the indicators found from the survey data. The criteria used for the factor analysis are (a) the eigenvalue is larger than 1, (b) the loading coefficient is larger than .40, and (c) there is no common loading on two or more factors. Five factors found from faculty perceptions of teaching digital technology are able to form new variables. Each new variable is the sum of its indicators. And the reliability and validity of each variable are checked with its Cronbach reliability coefficient (() and correlation coefficients among its indicators (concurrent validity). The valid variables derived from factor analyses are listed as follows.

Change (changes brought by teaching digital technology, including 5 indicators, eigenvalue = 4.708, percentage of variance = 18.1%, (= .79, n = 66)

Digital production

1. Is more time-consuming
2. Is more difficult for instructors to teach students
3. Is easier for students to learn
4. Requires more individualized instruction
5. Requires more assistance for students in the lab

Support (needs for budget and technical support, including 5 indicators, eigenvalue = 2.637, percentage of variance = 10.1%, $\alpha = .74$, n = 67)

1. Digital production is more difficult for instructors to handle technical problems
2. System upgrades of digital production are too expensive
3. The school did not have enough funding for purchasing digital systems
4. The administrator did not support adopting digital systems
5. There was no technical support

Learn (including 2 indicators, eigenvalue = 2.317, percentage of variance = 8.9%, $\alpha = .77$, n = 68)

1. Digital production is easier for me to learn
2. Digital production is easier to operate

Function (accessibility and impact on student learning, including 4 indicators, eigenvalue = 1.537, percentage of variance = 5.9%, $\alpha = .68$, n = 66)

1. Digital production is more accessible to students
2. The price is too high to afford a digital system
3. Digital technology more readily facilitates students' creativity
4. Digital technology better stimulates students' curiosity

Confiden (confidence of using digital technology, including 2 indicators, eigenvalue = 1.106, percentage of variance = 4.3%, $\alpha = .67$, n = 72)

1. I do not know very much about digital systems
2. I do not feel comfortable to use digital system

There are 2 factors found from faculty's perceptions of students' attitude toward digital technology. They are listed as follows.

Attitudp (positive attitude, including 2 indicators, eigenvalue = 2.562, percentage of variance = 36.6%, $\alpha = .82$, n = 68)

1. Students are eager to learn digital technology
2. Students are curious to explore digital technology

Attitudn (negative attitude, including 3 indicators, eigenvalue = 1.018, percentage of variance = 14.5%, $\alpha = .69$, n = 65)

1. Students are afraid of operating digital technology
2. Students are frustrated when encountering problems
3. Students are confused by digital technology

There are 4 factors found from students' perceptions of learning digital technology. These variables are listed as follows.

Learn (using and learning digital technology, including 6 indicators, eigenvalue = 5.417, percentage of variance = 20.8%, $\alpha = .88$, n = 192)
Digital production

1. Is more productive
2. Is more accessible to students
3. Is easier for me to learn
4. Is easier to operate
5. Is easier for students to learn
6. Is easier to set up for working on my projects

Confiden (confidence of digital system quality and usage, including 6 indicators, eigenvalue = 3.261, percentage of variance = 12.5%, $\alpha = .82$, n = 188)

1. I do not know very much about digital systems
2. There was no technical support
3. There was no formal training for using digital systems
4. The system did not work very well
5. I do not feel comfortable using digital system
6. The quality of the systems is not good enough

Support (budgeting for digital systems, including 4 indicators, eigenvalue = 1.923, percentage of variance = 7.4%, $\alpha = .73$, n = 183)

1. Digital production is too expensive to upgrade the system
2. The school did not have enough funding for purchasing digital systems
3. The administrator did not support adopting digital systems
4. The price is too high to afford a digital system

Function (impact on student learning, including 2 indicators, eigenvalue = 1.582, percentage of variance = 6.1%, $\alpha = .82$, n = 203)

1. Digital technology more readily facilitates students' creativity
2. Digital technology better stimulates students' curiosity

There are 2 factors found from student self-reports of their attitude toward digital technology. These are listed as follows.

Attitudp (positive attitude, including 3 indicators, eigenvalue = 3.711, percentage of variance = 53%, $\alpha = .85$, n = 141)

1. I am eager to learn digital technology
2. I am curious to explore digital technology
3. I am optimistic about the rapid development of digital technology

Attitudn (negative attitude, including 3 indicators, eigenvalue = 1.406, percentage of variance = 20.1%, $\alpha = .77$, n = 141)

1. I am afraid of operating digital technology
2. I am frustrated when encountering problems
3. I am confused by digital technology

Student positive attitudes toward digital technology include "eager to learn," "curious to explore," and "optimistic toward digital technologies."

Table 7.4
Product Moment Correlation between Students' Perceptions of Digital Technology and Students' Attitude Toward Digital Technology

	Attitudp	Attitudn	Learn	Support	Confiden	Function
Attitudp	1.000					
Attitudn	– .402***	1.000				
Learn	.554***	– .376***	1.000			
Support	.179*	– .012	.059	1.000		
Confiden	– .123	.263**	– .237**	.281**	1.000	
Function	.563***	– .172	.521***	.147	– .169	1.000

Note: N = 126, Missing = listwise
* p < .05, ** p < .01, *** p < .001

The results show that student positive attitude (Attitudp) is significantly and positively related to their learning experience (Learn) (n = 126, $r = .554$, $p < .001$), understanding of the current funding and supporting problems (Support) (n = 126, $r = .179$, $p < .05$), and appreciation for "facilitating students' creativity" and "stimulating students' curiosity" by digital technologies (Function) (n = 126, $r = .563$, $p < .001$). Student negative attitudes toward digital technology (Attitudn) consist of "afraid of operating digital technology," "frustrated when encountering problems," and "confused by digital technology." The results show that student negative attitude (Attitudn) is significantly and negatively related to their learning experience (Learn) (n = 126, $r = -.376$, $p < .001$) and positively related to their lack of confidence in learning digital technologies (Confiden) (n = 126, $r = .263$, $p < .01$), such as "lack of knowledge about digital systems," "lack of technical support," "lack of formal training," "system's not functioning well," and "not feeling comfortable to use digital technology." Student learning experience (Learn) is significantly and negatively related to students' lack of confidence (Confiden) (n = 126, $r = -.237$, $p < .01$) and positively related to student appreciation of digital technology (Function) (n = 126, $r = .521$, $p < .001$). Moreover, student perceptions of the lacking support (Support) is significantly and positively related to their perceptions of lacking confidence with using digital technologies (Confiden) (n = 126, $r = .281$, $p < .01$).

In Table 7.5, there are only a few significant correlation coefficients. Faculty perception of the changes needed for teaching digital technology

Table 7.5
Product Moment Correlation between Faculty Perceptions of Digital Technology and Students' Attitude Toward Digital Technology

	Change	Support	Attitudp	Attitudn	Learn	Function	Confiden
Change	1.000						
Support	.180	1.000					
Attitudp	– .017	– .037	1.000				
Attitudn	.505***	.256	– .237	1.000			
Learn	– .365**	– .112	– .135	– .271*	1.000		
Function	– .237	– .058	.173	– .373**	.167	1.000	
Confiden	– .050	.214	– .110	.166	– .139	– .088	1.000

Note: N = 58, Missing = listwise
* p < .05, ** p < .01, *** p < .001

(Change), consisting of "more time-consuming," "more difficult to teach," "too expensive," "not easier for students to learn" (recoded from its original scales), "requiring more individualized instruction" and "requiring more assistance for students in the lab," is significantly and positively related to their understanding of student negative attitudes toward digital technology (Attitudn) (n = 58, $r = .505$, $p < .001$). The more instructors know about student problems with digital technology, the more they feel the necessity to work with students in order to help students learn. On the other hand, faculty perception of the changes (Change) is significantly and negatively related to their own learning process (Learn) (n = 58, $r = -.365$, $p < .01$). The easier a faculty member learns digital technology, the less change he or she feels necessary to make in his or her teaching. Further, faculty's learning experience (Learn) is significantly and negatively related to their understanding of student negative learning attitudes (Attitudn) (n = 58, $r = -.271$, $p < .05$). The easier an instructor learns digital technologies, the less he/she understands student problems with learning digital technology, such as "afraid," "frustrated," or "confused." Faculty's perception of student negative attitude toward digital technologies (Attitudn) is significantly and negatively related to his or her perceived functions of digital technology (Function) (n = 58, $r = -.373$, $p < .01$), such as "more accessible," "facilitating students' creativity," "stimulating students' curiosity," and "affordability of digital systems" (recoded from the original scales of "price too high").

QUALITATIVE RESULTS

Student comments

Many student respondents express their desires for future jobs in electronic media industries, such as TV, radio, newspaper, web page design, production house, movie studio, and so forth. These students want to learn digital technology in college in order to get a good job and earn a good salary. Their typical comments are as follows:

I plan to do some sort of web page design for any company that'll be kind enough to hire me. I expect digital technologies to continue to improve to the point that there will be no loss in quality at any point in the editing process.

Students need to be trained in digital technology because that is where the industry is going. We need time with and access to the equipment. I want to learn all of the digital technology I can. I think the new technology has been great in making things simple but the skills to do these things are very important.

I want to work at a production house or in the productions department of an existing corporation. I'm sure that digital technologies will develop rapidly in the future and in a few years, everyone will be using digital technology and creating better and more advanced projects because of it.

Having access to all the current and updated technologies is a must in order to keep up with the rapidly changing world.

In the few classes I've had dealing with digital technologies, I haven't had much of a chance to learn what I'd like to about all the equipment. I would like to feel more comfortable around video and audio equipment, but right now I really don't.

I hope to work in the television or movie-making field. I will need to learn the technology but I think it will be on-the-job training because the programs we use here can give me a foundation but I think each job will use different technology.

If we are going to teach the major we should give students an education for tomorrow not yesterday. Digital technology will become cheaper, but only as newer and better technology comes into existence. We need to keep the college on the cutting edge, not continuously hoping to get a sharpener for the old stuff we already have.

I wish that we had more individual instruction for digital technologies. Right now, we are pretty much forced to learn it on our own. That makes it very hard to complete assignments in a timely manner, and since we don't know how to work the darn stuff, we often just figure out how to perform the basic operations, and don't know what other wonderful advanced stuff we can do to enhance our productions.

I just wish that instructors were given a formal training so that they understand it also, allowing us to be trained in digital operations the correct way. I wish that our school would shell out a little more money to keep up with the ever-changing technology so that students can stay current in what they are learning.

Faculty Comments

What concerns faculty respondents the most is the funding for upgrading digital systems, technical support, formal training opportunities for learning up-to-date digital technologies, the change in teaching-learning relationships, and the marketable skills possessed by electronic media graduates. Some opinions are shown as follows:

I found two problems with digital applications: (1) Students can work the computer, but that doesn't mean they can do production; (2) it is at times apparent that students learn "production" aspects better in the analog domain, as compared to digital. Here, the ideas and concepts of mixing, layering, and basic relationships of components seem to come together more applied in the analog world. Maybe it is because the student is forced to work the various pieces of equipment and mediums (i.e., tape) in the analog systems.

[It] depends on the type of production and time constraints. Creatively, digital [system] is best for sports, and [for] other live action or material that must be cut simply and quickly—analog is better.

[Digital technology provides a] wide range of knowledge for students, some know a lot and others have little or no knowledge.

I think the learning curve is slow, but once you really know the Media 100, it is a splendid teaching tool. The students learn faster than the faculty. But the faculty absolutely needs to have major tech support, which is often not available. (We have NONE in the dept, but the media center where the undergrads work does have good tech support. However, for the grad students and faculty, absolutely zero.)

It needs more individual instruction AT FIRST, . . . then they are off on their own, but they still need tech support.

The learning curve on digital equipment is steep at first, but is worth the effort in the long run, because students are increasingly using the technology in the workplace, and a student trained in digital production is more marketable. Our school does not provide any assistance for instructors wishing to use digital system, audio or video. All the learning by instructors is done on their own time and expense.

We are extremely fortunate to have very good support for digital production and we currently have one of the best-equipped digital video production labs in the United States.

Although digital equipment is good to know, it is still a matter of individual creativity. A student can have access to all of the "bells and whistles" but still end up with bad projects. . . . Digital equipment does not make students more creative; it simply makes it easier and more efficient for students to accomplish certain tasks. I have had some troubles with digital. It seems to make the hard stuff easier to do and the easy stuff harder to do for some reason.

DISCUSSION: PERCEPTIONS OF DIGITAL SYSTEMS—THE GENERATION GAP

Since the student respondents were enrolled in college education in 1995 or later, they have been studying in the computer environment in college, or even in high school. Generally speaking, they seem to be adapting to the information age, including digital technology, better than their instructors' generation that was trained in analog systems. Therefore, it is not surprising that students perceive more advantages of digital technology to their learning and less problems with using digital systems, such as "time consuming" and "difficulty to teach digital technologies," than instructors, shown in Table 7.1. While faculty members complain about the lack of formal training, students do not have the similar experience because 50% of the students learn digital technology by taking courses. Although the mean scores in Table 7.1 show that students agree that learning digital technology needs more individualized instruction and assistance in the lab, their perceptions are significantly lower than instructors' perceptions. The mean scores in Table 7.2 indicate that students do not feel the lack of digital skills and knowledge, nor do they feel uncomfortable using digital technologies, though they may not feel as much confidence as their instructors in using digital systems. However, students are aware of the increasing demand for digital technology in the job market and are eager to prepare themselves to be more competitive in the marketplace. Therefore, they are more sensitive to the support from the administrators, the upgrading prices, and the quality of the systems than instructors, shown in Table 7.2.

The survey results are consistent with the findings from a survey of WebCT in 2000 by the author. Students in communication technology classes answered the questions about their attitude toward new technologies, possession of computers, access to the Internet, and the comparison of teaching-learning relationships between the WebCT-oriented classroom and the traditional classroom. The results show a higher percentage of college student computer ownership and access to the Internet services, shown in Table 7.6. The majority of the students express their desire to learn and feel comfortable with new technology. These students also appreciate the interactive teaching format that allows them to turn in assignments through WebCT. However, they do not think the WebCT assisted classroom

Table 7.6
Student Computer Ownership and Internet Usage (%)

	Yes	Going to buy one	Not decided	Have none	Don't want
Students have personal computers	68.5	7.4	3.7	3.7	20.4
Parents have computers	88.9	0	1.85	9.3	
	4+	3	2	1	0
Families have computers	27.8	18.5	31.5	16.7	5.6
	Daily	**Weekdays**	**2-3 times per week**	**Once per week**	**<1**
Visit to WWW	83.3	7.4	7.4	1.85	0
	Cable modem	**Telephone Modem**	**Free access**	**Going to**	**None**
Subscription to the Internet service	31.5	13	35.2	5.6	14.8

Note: N = 36

teaching has changed the teaching-learning relationship very much. This fact indicates that students are excited by using digital technology but are still not aware of their new role in the teaching-learning relationship required by the paradigm change in the digital age.

Obviously, the gap between students and instructors occurs within the social context of the digital age. Instructors in electronic media education have no time to wait for the changes. Rather, they need to actively move forward to meet the challenge.

THE PARADIGMATIC CHANGE OF TEACHING-LEARNING RELATIONSHIP

Teaching digital technology is an ongoing process. Electronic media instructors need to learn new knowledge and skills constantly in order to teach students up-to-date digital technology. The survey results and faculty comments indicate that the instructors' learning experience (Learn) helps them better understand students' learning problems (Attitudn), and their awareness of students' problems helps instructors explore new teaching strategies (Change), shown in Table 7.5. On the other hand, students' learning desire (Attitudp) enhances their awareness of the problems that instructors encounter (Support), shown in Table 7.4. Faculty comments show that instructors are concerned about the relationship between digital skills, creativity, and the knowledge gap among students. Some believe that students need to be able to use both digital and analog production processes, depending on the types of tasks. They do not want students to overuse digital transitions and effects that may replace student creativity. Rather, students need to pay more attention to content and aesthetics in production.

This study depicts a beginning stage in the transition of electronic media education from analog to digital technologies. Digital technology has created a new teaching-learning environment that facilitates the shift of the teaching paradigm from teaching-centered to learning-centered. It promotes cooperative, collaborative, and supportive learning. Students are responsible for their own learning. However, at the beginning stage, instructors are struggling to gain confidence and convenience, and to reduce cost in their own learning process, as well as to enhance student access, quality, and productivity. Students are expecting to learn the up-to-date technologies as much as possible at the college level and to find a good job in the industry after graduation. With the push to participate in the technological innovation, instructors do not have time to think systematically about the paradigm change.

Nonetheless, there are demands for change. For instance, the first factor found in the faculty survey is Change that consists of "more time consuming," "more difficult to teach," "too expensive," "not easier for students to learn" (recoded from its original scales), "requiring more individualized instruction," and "requiring more assistance for students in the lab." Some faculty members comment that the learning curve is slow but students are faster learners. Some students complain that they learned a lot on their own and have not learned all the skills they would like to learn. The results support Fink's (1998) findings. The instructor's role is no longer a disseminator of knowledge and skills but instead an interactive facilitator who initiates, guides, and oversees the interactions of student with instructor and student with other students. Students need to understand that learning must be a process of discovery through cooperative efforts with others and, therefore, take responsibility for continuously expanding their own knowledge and skills by engaging more in self-education under his or her instructor's supervision (Fink, 1998). Now is the time to actively change from the Instruction Paradigm to the Learning Paradigm in electronic media education.

CONCLUSION

This is an exploratory study about the changes in teaching-learning relationships in electronic media education. It collected robust information about electronic media, faculty and student perceptions of digital technology, the problems with implementation of digital systems, student attitudes toward digital technology, and the factors that impact on student learning. The results demonstrate students' enthusiasm in learning digital technologies and their high expectation of the impact of learning digital technologies on their future job opportunities. This study indicates that instructors are adapting themselves to the digital age but some are having difficulties finding the administrative support to upgrade their knowledge and skills. It also shows that instructors are aware of the differences between teaching digital and analog technologies and that they are anticipating the paradigmatic change in the teaching-learning relationship.

There are limitations to this study. One is a methodological problem. It was very difficult to gather information from electronic media faculty nationwide. Although the online survey was posted on the Internet for about four months, and the survey was announced through various listserv groups, there were only 72 responses. There may be reasons. First, the questionnaire included many technological terms and product brands that may confuse the beginning adopters. Second, the online survey may not be a format for beginners to use conveniently. Therefore, the finding of the study is not intended to be generalized to all the schools that offer electronic media majors.

Third, there is a shortage of theoretical references on this topic. Most of the existing studies referred to were related to the field of education. This reminds communication scholars to do more research on this topic, especially regarding the interactions between electronic media instructors and students. A complementary qualitative study may be beneficial for better understanding the paradigm changes underway in electronic media education.

REFERENCES

Barr, R.B., & Tagg, J. (November/December, 1995). From teaching to learning: A new paradigm for undergraduate education. *Change*, 13–25.

Bruffee, K.A. (1999). *Collaborative learning: In High education, interdependence, and the authority of knowledge* (2nd ed.). Baltimore, MD: The John Hopkins University Press.

Brunner, C., & Tally, W. (1999). *The new media literacy handbook: An educator's guide to bringing new media into the classroom.* New York: An Anchor Book, Doubleday.

Chen, Z. & Bankston, R. (1999). *The diffusion of computer skills in communication curricula: Is there a gap between the educational experience and employers' needs?* (ERIC Document Reproduction Service No. ED 420 886)

Connick, G., & Russo, J. (1995). Technology and the inevitability of educational transformation. In E. Boschmann, (ed.). *The electronic classroom: A handbook for education in the electronic environment.* (14-20). Medford, NJ: Learned Information, Inc.

Fink, E. J., (1998, April). The transistion to digital video: What lessons have we learned so far? *Paper presented at the annual meeting of the Broadcast Education Association.* Las Vegas, NV.

Green, K.C., & Gilbert, S.W. (March/April, 1995). Great expectations: Content, communications, productivity, and the role of information technology in higher education. *Change,* 8–18.

Kozma, R., and Johnston, J. (January/February, 1992). The computer revolution comes to the classroom. *Change,*10–23.

Lind, R. A., and Danowski, J. A. (1998). Training tomorrow's broadcasters: A comparison of employers' needs and educators intentions (NAB Research Event 1998-1999 Report). Retrieved July 30, 2003 from www.tigger.vic.edu/~rebecca/

Regan, S. (1995). Teaching visual analysis using CAV interactive videodisk technology. In E. Boschmann, (ed.). *The electronic classroom: A handbook for education in the electronic environment.* (130–137). Medford, NJ: Learned Information, Inc.

Steele, R. L. (1995). Pedagogical issues. In E. Boschmann, (ed.). *The electronic classroom: A handbook for education in the electronic environment.* (21–29). Medford, NJ: Learned Information, Inc.

Van Tassel, J. M. (2001). *Digital TV over broadband: Harvesting bandwidth.* Boston: Focal Press.

CHAPTER 8

Outcome Differences in Modular Versus Traditional Instruction in Audio Production Assignments

Joseph R. Blaney, Gerard Donnelly, and Matt Rouch

New technologies have improved the ease and efficiency of offering college course work with nontraditional methods. In fact, some institutions have embraced such learning processes so strongly that degrees can be obtained with little or no time spent in traditional campus environments. Moreover, even students physically present on university campuses are availing themselves of modular learning opportunities.

Of course, the existence of such technologies raises the perennial question of whether something ought to be done simply because it is possible. Should we fry bacon in a toaster simply because the technology allows it?

As mass communication programs (including the program at Northwest Missouri State University, institutional home of authors Donnelly and Rouch) are asked to develop distance and other modular course offerings, it is worth considering whether students will enjoy equality of outcome from these differing pedagogical situations.

As such, this study asks:

1. Are there differences between traditional and modular learners in broadcast production course sequences?
2. Are there differences between the genders in the caliber of broadcast production assignments?
3. How do students characterize the learning process in modular versus traditional formats?

For purposes of this study, "modular" learning is defined as the asynchronous presentation, consumption, and completion of instructional content and assignments. We use the term "modular," as opposed to "distance," because it is more inclusive. Distance education is largely conceptualized as course delivery via correspondence and/or technological facilitations like interactive television and the World Wide Web. Modular learning can be applied to any learning arrangement wherein completion of course readings, assignments, and activities occurs at the student's pace, without physical proximity to the instructor.

METHOD

An Introduction to Broadcast Operations course at a regional, midwestern, state university provided an excellent opportunity to experiment with students who had no previous broadcast production experience. Four sections of this course created a sample size of 73 subjects, 38 women and 35 men. All four sections proceeded normally in offering course content with a minor variation. For the course sequence dealing with a basic radio production assignment, the sections were divided into two "traditional" groups and two "modular" groups. This allowed us to divide the sample roughly in half. Thirty-eight students were given traditional instruction and thirty-five students were given modular instruction.

The assignment was to produce a basic, 30-second voice/music/sound effects radio commercial employing three respective stereo tracks. The facilities used included two identical Digidesign Pro-Tools LE digital audio workstations. Each assignment was rated according to four factors: overall levels, mixing levels, editing, and general aesthetics.

Two independent raters (two of the three authors) rated 20% of all assignments on a one to five scale for each factor, with one indicating lowest caliber and five indicating highest caliber. Positive and significant agreement was achieved for overall levels [$r(d.f. = 71) = .84$, $p<.01$], mixing levels [$r(d.f. = 71) = .82$, $p<.01$], editing [$r(d.f. = 71) = .87$, $p<.01$], and general aesthetics [$r(d.f. = 71) = .84$, $p<.01$]. In short, reliable measurement of each of these factors was achieved.

This data allowed us to run two different types of *t*-tests. First, we looked for differences between modular learners and traditional learners in caliber levels for each factor. Second, we assessed differences in caliber between men and women in both traditional and modular instructional environments. These *t*-tests allowed us to answer our questions about differences between the two types of instruction at issue while also offering a glimpse into any existing gender differences.

In order to answer our question about how students describe the modular learning process, focus groups were conducted and that data was analyzed with a grounded theory (Glaser and Strauss, 1967) approach. As such, this study worked from the strength of mixed methodologies. As Tashakkori and Teddlie (1998) noted, this allows researchers to combine generalizable claims with a more rich and subtle understanding of what the data mean.

RESULTS

Quantitative Analysis

Universally, we found no significant differences between the modes of instruction or the genders for any of the factors. Specifically, there were no significant differences between modular and traditional groups for overall levels [t(d.f. = 71) =.008, p>.05], mixing levels [t(d.f. = 71) = .276, p>.05], editing [t(d.f. = 71) = .733, p>.05], or aesthetics [t(d.f = 71) = 1.233, p>.05]. In short, student performance on the radio assignment did not differ between the modular students and traditional students.

Moreover, we found no differences between the genders for any of the factors. There were no significant differences between men and women for overall levels [t(d.f. = 71) = .181, p>.05], mixing levels [t(d.f. = 71) = .516, p>.05], editing [t(d.f. = 71)=.442, p>.05], or aesthetics [t(d.f. = 71) = .252, p>.05]. Again, the completion of the assignment did not produce differences between men and women.

The quantitative data presented here answers our first two research questions. Namely, there are no differences between traditional and modular learners in a simple broadcast production assignment. Neither are there are differences between genders in such assignments. However, our third research question, assessing students' characterizations of modular versus traditional learning processes, required a more rich observation and explanation. This "thicker description" follows in the next section.

Qualitative Analysis

In order to add a qualitative dimension to the study of outcome differences, in-depth interviews were conducted with 15 of the student participants. The purpose of the interviews was to gather as much information as possible about the students' experiences as they completed the audio production assignment without the instructor present. The quality of the final projects would tell us how successful they had been at learning about and using the Pro-Tools digital system in a modular setting. However, only by asking questions and listening to the students' comments could we know how those students perceived the modular experience in terms of the quality of instructional delivery.

The methodology utilized was Glaser and Strauss's (1967) "constant comparative" approach in which researchers gather data and seek to define relationships among the data. As initial data is generated (by listening to focus group interviewees), it is assigned to separate categories. As more data is gathered, the categories become increasingly distinct from one another. Finally, those categories solidify to the point that new data becomes unlikely to alter their defining characteristics. As the interviews were conducted, this approach made it possible to identify patterns in the students' responses and to draw some inferences about the effectiveness of the modular assignment from a pedagogical perspective.

The first pattern to emerge had to do with the students' initial reactions when learning about the modular assignment. Although participation was voluntary, nearly all expressed feelings of apprehension about their prospects of completing the project in a satisfactory manner without the instructor present to answer questions in the studio. One student said, "I came from a really small high school and never even used a Mac before. When I found out that I had to do the assignment without the instructor, I was pretty intimidated." Other students expressed similar anxieties about the project. Several noted that there was a general tendency among many of the students to procrastinate and put off the project because they were worried about their ability to complete it.

A second pattern that emerged during the interviews centered on the amount of contacts necessary with the instructor via e-mail. More than half the students reported e-mailing the instructor at least twice for clarification about the instructions. One student observed that there were so many buttons and switches on the Pro-Tools edit screen and mixer screen that she easily became confused. "I needed to ask a lot of questions and follow-up questions early to get started," she said. "There were just too many buttons. It was mostly basic things that I needed to get started. I needed help finding the right channel inputs because I wasn't sure where they were located on the screen."

A third pattern in the students' responses was concern about the amount of time that it took to finish the project. Nearly all the students remarked that the assignment took much longer than they had anticipated. "I must have spent 10 hours in the studio trying to produce a 30-second commercial," said one student. "When we got the assignment, I thought it would take a couple of hours at the most." Another student mentioned that part of the delay was caused by having to wait for the instructor to clarify a step by returning an e-mail message before being able to continue on with the project.

"I would have the studio reserved," he said, "and not be able to finish the work. So basically I would be starting over again the next day."

These concerns would appear to suggest a degree of frustration and even dissatisfaction with the modular experience. However, as the interviews continued the vast majority of the students reported actually coming away from the assignment with a very positive impression. To some extent, this impression was the result of feeling that they had spent enough time working with the digital editing system to become truly knowledgeable about it. Even though it took longer for most to begin the work and to complete it, the students saw benefits from the amount of hands-on experimenting they had engaged in. "The assignment really forced us to get in the studio and spend time on the Pro-Tools system," said one student. "In my classes in the past the equipment was demonstrated for us but it's hard to learn by watching someone else operate it. This much hands-on experience made me feel that I knew how to get going with projects. The written instructions were also very helpful because you could refer back to them if you forgot a step."

Another student observed that the modular experience was less stressful than a regular lab with an instructor and a time limit. "You could stay and work with the system for as long as you wanted," she said. "You could experiment and make mistakes and eventually figure it out without worrying about how everything you do might affect your grade." Several students made similar observations about the relative lack of pressure in an individualized setting and the opportunity to learn the editing system at one's own pace. "There were times when I needed some help," said one. "But for the most part I was able to navigate the system. When I was done I felt proud that I had learned it by myself." In the final analysis, those comments might be the most telling ones about the benefits of the modular experience. The individual, hands-on setting provides students the opportunity to learn the basics of a digital editing system at their own pace, to keep exploring even after the required assignment has been completed, and eventually to become adept at the finer points of the system.

CONCLUSION AND INSTRUCTOR'S COMMENTS

Following are some observations from the instructor of the course about how students seemed to react to the assignment differently, depending on whether they were part of the traditional or modular group.

First of all, in both groups many students waited until there were only a few days left to get the assignment done before they did anything at all. This is not surprising, but students in the modular groups seemed far more likely to put off the project even though they had far more time to get it done. Students in the traditional group typically had three to four weeks to get the assignment done; students in the modular group had almost the entire semester. About one-third of the students in the traditional group waited until the last week to get started; *at least* half of the students in the modular group put things off until the last week. The instructor was able to determine this, by the way, by looking at the sign-up sheets for the studios.

I believe the reason may have been due to a difference in perception about the value of the assignment as well as the level of difficulty. The traditional group actually saw the instructor put the entire commercial together with Pro-Tools projected onto a big screen. Many seemed to think it looked like an interesting and fun skill to learn. By contrast, based on conversations and anecdotal comments, many in the modular group seemed to dread having to read the step-by-step instructions and to perform the task by themselves.

Second, it is clear that the modular group actually did have to work harder and more hours to complete the assignment. As a rough estimate, I would say (again, looking at the studio sign-up sheets) that the modular group put in close to seven or eight hours on average while the traditional group put in an average of perhaps three or four. Of course, the sign-up sheets do not tell you if the students were actually *in* the studio doing the job, only that they had signed up for the time.

Third, the traditional group was far more likely to take advantage of their ability to directly ask the instructor questions. They would often simply stop by the instructor's office while they were working in the studio if a question popped up. The modular group, limited to e-mail, asked virtually no questions whatsoever. There were perhaps five or six e-mail questions.

Finally, the problems encountered by students in the modular group seemed quite often to be the result of not reading the instructions closely enough. Often they would try to rush through the assignment as quickly as possible, missing important points—and not learning from their mistakes. In many instances the irony was that rushing probably caused them to have to spend several more hours on the project than if they had simply relaxed and read the instructions closely.

In fairness to modular learning, however, this is probably the result of maturity and is not an inherent weakness of the modular approach. The majority of the students were freshmen and sophomores, in their late teens, and many had not yet cultivated the virtue of patience.

In conclusion, there appears to be both long-term potential and immediate problems with the modular form of instruction when it comes to an audio recording assignment. If students take the time to read the instructions carefully and to e-mail appropriate questions, it is clear that most are fully capable of learning a software editing system such as Pro-Tools and completing projects of acceptable quality. Over time and with additional practice, their adeptness with the editing system will no doubt continue to improve. In the short term, however, learning such a system requires an additional investment of time that many students might be reluctant to make.

REFERENCES

Glaser, B. G., and Strauss, A. L. (1967). *The discovery of grounded theory: Strategies for qualitative research.* Chicago: Aldine.

Tashakkori, A., & Teddlie, C. (1998). *Mixed methodology: Combining qualitative and quantitative research.* Thousand Oaks, CA: Sage.

CHAPTER 9

Legal Issues and Digital Technology in Broadcast Education

Dom Caristi

This chapter is likely to be the most contentious and aggravating of this entire volume. While faculty in electronic media are often interested in ways in which they can expand their use of digital media in classrooms, they are less interested in hearing of the legal ramifications that may limit their use of educational materials. Without question, the single greatest limiting factor of all is copyright and its impact on the use of sounds, pictures, and moving images that belong to others.

As a body of law, copyright is as old as our nation itself. The U.S. Constitution empowers Congress to grant exclusive rights for a limited time to authors and inventors. The extensive rules created by Congress in the various Copyright Acts in the past century have been attempts to draw the line between the rights of the authors and the free exchange of information. There are extensive volumes on copyright (Fishman, 2000; Goldstein, 1999; Halpern et al., 1999) that provide much more detail. The U. S. Copyright Office maintains a Web site (*http://lcweb.loc.gov/copyright*) with detailed information and links to dozens of publications. In spite of all this available information, many media educators are confused about the applicability of copyright law to what they do. A 1997 survey of broadcast educators demonstrated that among those who teach courses in broadcasting there is still a lack of basic information (Braun, 1997). Respondents to a true/false test of copyright issues related to educational use answered correctly only 67% of the time. Clearly such a percentage suggests a lack of understanding about the applicable copyright laws for broadcast educators. The disparity among educators was also significant. The individual respondents' scores ranged from 88% correct to only 25% correct. The purpose here is not to explore every nuance, but instead to inform the reader of the applicability of copyright laws to broadcast education.

Perhaps copyright is violated so often because it is so much easier to do than it ever has been. Omnipresent photocopiers make it possible to legally copy a page from a magazine or to illegally photocopy an entire best-selling book. Videocassette recorders are located in the majority of American homes and classrooms and they make taping of programs relatively easy. Movies can be rented on tape or DVD not just at large video store chains but even at local gas stations. The Internet now makes it possible to download digital files of pictures, registered trademarks, sound files, and even motion pictures. Violating copyright is as easy as clicking a mouse in the privacy of your home.

Perhaps because online copying is so simple to do, so many people think it's acceptable to copy anything from anywhere and use it indiscriminately. It may be tempting to see some graphic or image at a Web site that would be useful and copy it onto your Web site. While that is sometimes true, it is not always the case. Certainly content creators who have a strong commercial interest in their images want to protect their financial interests. Cartoon characters have great commercial value. In fact, many of the companies that license the characters will sell their images for use in slide shows, on Web sites, and in multimedia presentations. Users should not assume that because something is available online, it is therefore license-free.

> Educators and students are advised to exercise caution in using digital material downloaded from the Internet in producing their own educational multimedia projects, because there is a mix of works protected by copyright and works in the public domain on the network. Access to works on the Internet does not automatically mean that these can be reproduced and reused without permission or royalty payment and, furthermore, some copyrighted works may have been posted to the Internet without authorization of the copyright holder (Consortium of College and University Media Centers, 2001).

Even Web sites that do provide free images are moving more toward "click-wrap" agreements. In software, the term "shrink-wrap licensing" applies to the notion that by removing the software from the packaging, the purchaser has agreed to the licensing conditions stated on the box. Click-wrap licensing is the extension of that concept to cyberspace, where individuals downloading images and software must first click on a screen of conditions and agree to them before receiving the file. At least one court has upheld the enforceability of click-wrap agreements.[1]

Copyright laws are still valid regardless of the number of violations. I like to use the metaphor of running a red light at 3:00 A.M. Is it still illegal? Most certainly. Can the perpetrator expect to be caught? Highly unlikely. Still, prosecutions do occur. The distribution of publicly televised programs to schools by a not-for-profit corporation was determined to be a copyright infringement,[2] as was the copying of 11 pages on cake decorating by an adult education teacher for use in her own classes.[3] Copyright infringements by classroom teachers have traditionally been rather "small scale." A teacher who copies excessively usually distributes to a very small audience within a classroom and has previously gone unnoticed by authors and publishers protecting their copyrighted works. The same digital technology that makes pirating works so easy also makes their detection easier. Some content owners have begun surfing the Internet to find unauthorized uses of their copyright-protected material. Businesses have been created that work as "cyber bounty hunters" (Stim, 2000). Where once it was impossible to investigate thousands of classrooms for copyright violations, technologically it is possible to visit millions of Web pages seeking unauthorized uses.

Professors ought to educate their students about copyright law and enforce the rules, not just because of the increased likelihood of prosecution, but because it is the right thing to do. Students who would never dream of stealing money or property have admitted to plagiarism and other forms of cheating, often rationalizing that "no one is hurt" or that "everyone does it." Higher education should demand that students adhere to the rules of ethical and legal behavior and not simply allow students to violate copyright laws because their institutions cannot afford to pay the necessary rights fees, or because the processes for requesting permission are too time consuming. Perhaps it is worthwhile to remind students that they, too, will be content creators, and copyright laws that might limit their behavior now will work to their benefit in the future, both in terms of the creative control copyright gives them over their own work and the financial benefits copyright royalties provide.

When the professorate uses copyrighted material without permission from or compensation to the copyright holder, the assertion is usually that the use of copyrighted materials for educational purposes is a *fair use.* It is necessary to discuss the details behind the Doctrine of Fair Use.

DOCTRINE OF FAIR USE

The specific rules regarding fair use have been developed through congressional action, advisory council recommendations and court decisions primarily from the last 30 years, but the concept of fair use has been around much longer. In 1841 the Supreme Court first decided that the letters of George Washington could be used without the permission of the owner, provided the uses were limited (Jasper, 2000). Fair use is described in Section 107 of the Copyright Act, which is Title 17 of the U. S. Code. The relevant portion reads:

> [T]he fair use of a copyrighted work, . . . for purposes such as criticism, comment, news reporting, teaching (including multiple copies for classroom use), scholarship, or research, is not an infringement of copyright. In determining whether the

use made of a work in any particular case is a fair use the factors to be considered shall include—

(1) the purpose and character of the use, including whether such use is of a commercial nature or is for nonprofit educational purposes;
(2) the nature of the copyrighted work;
(3) the amount and substantiality of the portion used in relation to the copyrighted work as a whole; and
(4) the effect of the use upon the potential market for or value of the copyrighted work.[4]

The Doctrine of Fair Use requires that *all four* criteria be satisfied in order to qualify as a fair use. While every single use of copyrighted materials in a classroom satisfies the first criterion, unfortunately too often the analysis goes no further. Education is an appropriate "purpose," as outlined in the first part of the fair use analysis, as are research, news, commentary and criticism. We need not concern ourselves with the first criterion.

When considering the nature of the work being copied, courts have provided guidance as to which works are more appropriate for Fair Use. Nonfiction works are considered to be more appropriate than works of fiction. This is not to suggest that fiction works are not subject to fair use. One of the best examples of a fair use might be a commentary on a newly released motion picture that contains a brief clip about the movie. While the work is fiction, the limited use of the copyrighted work for the purpose of illustrating a criticism would be a protected fair use. Still, courts have stated that nonfiction works are more appropriate for fair use because of their use of facts that are more likely to be useful to those engaged in research and education.

Another factor to consider in examining the nature of the copyrighted work is the original author's desire to have the work widely disseminated. Again, courts have been clear that unpublished works are less suited to fair use protection than published works. An author who publishes a book has clearly sought public dissemination. On the other hand, a student who has been required to write an original poem for a class has not sought public attention and might be able to claim a copyright infringement if the poem is published without permission. Faculty members who post student work to the World Wide Web without obtaining their permission first are treading on dangerous territory.

The nature of the copyrighted work also implies that some works deserve more copyright protection than others do. It is a long-established rule that the simple statement of facts is not entitled to copyright protection. No one can copyright the facts contained in a news story, but the specific presentation of the story can be. It follows from this rationale that intellectual property that is more factually based is more inviting of fair use than material that is more creative in nature. The copying of news articles is more likely to be a fair use than the copying of a work of fiction. This should be seen as a continuum rather than a check-off for "fact-based" versus "creative" content.

The third criterion for consideration in fair use analysis is the amount and substantiality of the copyrighted work being used. This is both a qualitative and quantitative analysis. Copying an entire 400-page textbook is less likely to be a fair use than just copying a page from the same book. In fact, a detailed set of guidelines is provided for educators by the Library of Congress (Copyright Office, 1998). The guidelines provide information on the application of the Fair Use Doctrine for educational purposes. When it comes to copying prose works, the guidelines allow prose works to be copied up to

> either a complete article, story or essay of less than 2500 words, or an excerpt from any prose work of not more than 1000 words or 10% of the work, whichever is less, but in any event a minimum of 500 words.[5]

The guidelines are quite clear in the amount of copying permissible, but not quite as clear on the substantiality. The Supreme Court has made it clear that qualitative measures must also be considered when they ruled that excerpts from Gerald Ford's autobiography, though quantitatively small, went to "the heart of the book."[6] It is safe to assume that a movie review that contains a brief video clip of less than two minutes would be acceptable for fair use purposes, unless the clip reveals a surprise ending or gives away the identity of the perpetrator in a murder mystery.

Finally, the effect of the use on the *potential* market for the copyrighted work requires that a fair use not damage the author's ability to profit from the work, now or in the future. Federal courts have twice decided that copying services that provide "coursepacks" for students must pay copyright royalties.[7] This may be the criterion least understood by academics. It is not a fair use to copy for students the same chapter from a textbook each and every semester. While the first three fair use guidelines would be satisfied, repeated use demonstrates the centrality of that particular textbook to the course, and the need to compensate the author appropriately. This is also the criterion that limits off-air copying of television programs for classroom use and prohibits the use of consumer videos in classrooms.

The fair use guidelines stipulate that nonprofit educational institutions may record programs off-air for classroom use but only under strict conditions. Programs recorded off-air must be erased 45 days after their original record date. Teachers may only show the program in classes during the first 10 consecutive school days of the 45: beyond that, they may only keep the tape for teacher evaluation purposes, *not* to be shown again in classes. In devising these guidelines, a committee comprised of educators, copyright owners, and creative guild members attempted to strike a balance between the fair use of televised material by educators and the commercial interests of authors and publishers. Educators were seen as needing to have a way to show material of current interest without having to wade through a lengthy process of acquiring the rights, while copyright holders were recognized to have a great commercial interest in the long-term use of such videos. Undoubtedly high on the list of desirable content for classroom teachers is nonfiction content of the sort found on public television, the History Channel, and other networks carrying documentary works. Fair use guidelines are intended to make it possible for faculty to use a current video, but if their intention is to use the same video each semester for a course, it ceases to be a fair use. Under those circumstances, faculty would have time to contact the rights holder, obtain permission, or pay the appropriate licensing fee.

Perhaps one of the most frequent violations of fair use guidelines is when teachers visit their local video store to rent a tape of a movie for use in class. While the purpose of the use is educational in nature, the impact on the market is just as certain. Renting a Hollywood version of a Shakespeare play to show an English class deprives the screenwriter (not the playwright, whose copyright has expired) the compensation due for a public performance. Schools that follow the rules pay a much higher fee for renting movies for classroom use than those who break the law by showing videos licensed only for home use. Again, while it might seem unduly expensive, one's perspective changes when he or she is the one entitled to the royalties.

In a digital age, some of the rules of fair use need to be revisited. For example, it was never before a concern as to whether archival copies could be made of a work in case the original became damaged. No one considered such a use in the print era, but in a world of computer software, backup copies are the norm. A person who purchases a specific piece of software without making a copy runs a tremendous risk that the original will somehow be destroyed or corrupted, leaving the purchaser to lose hundreds of dollars. In a digital age, a single backup copy of a piece of software is an acceptable fair use. In the event the original is somehow damaged, the backup can be used. What is not acceptable is making multiple copies, or making a backup copy that is also in use. If multiple copies of software are needed, multiple copies (or a site license) should be purchased. Faculty should not purchase one copy of software for use in a class unless only one copy of that software is in use at a time. There is nothing illegal or immoral about purchasing only one copy and letting each student in the class use it, one at a time (which would be analogous to lending a book). Purchasing one copy of software and then making multiple copies for simultaneous use by multiple students is illegal.

The rules regarding classroom viewing of online materials is not quite as clear. What rules apply if a faculty member merely visits a Web site and plays an audio or video clip to the class from that site? Does this constitute a public performance, like the rental and projection of a video would? Instead, is it more like a classroom watching a television program in real time as it airs, which is perfectly acceptable under fair use guidelines? While there exists no definitive word from either the courts or Congress on this issue, copyright law exists primarily to protect the economic interests of the author, thereby encouraging further creative endeavor. Since the creators of Web pages make their content available to the public free of charge and invite wide dissemination by putting their work on the Web, it is unlikely that public display of Web sites in a class would trigger any copyright concerns. The situation might be different, however, if a Web site requires subscription. For example, only subscribers to the *Wall Street Journal* can access most of the content they provide online. A faculty member who paid an individual subscription to the *Journal* yet regularly projected that content for class use might be crossing the line between an acceptable fair use and a copyright infringement.

One other major impact the growth of the World Wide Web has had on educators and the use of copyrighted material is the extension of activity that was previously confined to the classroom to a worldwide audience. Many faculty members use the Internet to provide their syllabi online and by doing so make their materials available to anyone who wishes to access them. Technology exists to password protect those sites, but historically faculty do not take such steps.[8] As a result, anyone surfing the Net using a search term that would identify his or her specific content area could be directed to a class Web site. This serves to disseminate knowledge but it also can serve as a promotional tool for universities attempting to attract students. In fact, M.I.T. has gone so far as to announce it will make all of its courses available online by the year 2010 (Sappenfield, 2001). This includes not just syllabi, but also lecture notes, assignments, and reading lists: the intellectual property of the professors who teach the courses. Faculty members who put materials online ought to be aware that everything online is freely accessible unless password protected, and individuals who will not respect the authors' copyrights may access freely accessible material.

Faculty members who are particularly concerned about their intellectual property ought to take precautions to protect their online works. First and foremost is the need to provide copyright notice. While technically all work created in a fixed medium is copyrighted at the moment of creation (magnetic media such as computer files qualify as "fixed"), without copyright notice infringers can make the claim that they were unaware that the author had any interest in protecting the copyright. Some will argue that since the material was being made available freely to all, the author must not care. Those of us who create intellectual property realize that the wide dissemination of our work does not translate to disinterest in its use, but those who do not create intellectual property can sometimes be naïve about this fact.

While faculty must be careful to protect their copyright when publishing online, they must also be cautious not to infringe the copyrights of others. Because a much broader audience may view class Web sites, what were once acceptable fair uses in a confined classroom may not enjoy the same protection online. As the Copyright Office has stated, "Different rights are implicated by different educational activities, depending in part on the technologies used" (Copyright Office, 1999, p. 70). For example, if a professor were to show a videotape of a network television program in class within 10 days following its airing, the playing of the tape would be a fair use under the established guidelines. If, however, the professor digitized the same television program and made it available through the class Web site, even if only for 10 days, the use would not be fair at all. In the first place, dissemination of the video program could be potentially to millions of viewers (although most professors do not get quite so many hits on their class Web sites). Second, the ability of viewers to see the program more than once (which the guidelines do not permit) violates the guidelines established for taping programs for educational use.

These are precisely the concerns expressed by the Conference on Fair Use (CONFU) when dealing with the area of distance education. In trying to strike a balance between acceptable uses and protecting the rights of authors, CONFU attempted to provide the same sorts of fair uses permitted in a single classroom to also be acceptable for learning at a distance. The rationale behind this thinking is that whether students are in one location or many, fair uses ought to be analogous, and in many ways they are. In order to fall under such guidelines, the distance education program must be conducted by an educational institution (not a private company) over a secure network (ensuring that only those students enrolled in the class have access to the copyrighted materials). The one area that CONFU specifically stated it was *not* covering in its guidelines was asynchronous delivery of education.

> These guidelines do not cover asynchronous delivery of distance learning over a computer network, even one that is secure and capable of limiting access to students enrolled in the course through PIN or other identification system. Although the participants believe fair use of copyrighted works applies in some aspects of such instruction, they did not develop fair use guidelines to cover these situations because the area is so unsettled. The technology is rapidly developing, educational institutions are just beginning to experiment with these courses, and publishers and other creators of copyrighted works are in the early stages of developing materials and experimenting with marketing strategies for computer network delivery of distance learning materials. Thus, consideration of whether fair use guidelines are needed for asynchronous computer network delivery of

distance learning courses perhaps should be revisited in three to five years (Lehman, 1998, p. 48).

Other examples of copyright concerns that would be considered fair uses within the walls of a classroom but not once posted on the World Wide Web include the use of short excerpts from books, newspapers, and magazines, the displaying of copyrighted artworks, and the use of popular music for classroom examples. Fair use, which is intended to allow for *limited* copyright infringement, is not applicable when these infringements are given greater distribution via the World Wide Web. Playing a piece of popular music in class in order to discuss it would be a protected fair use, while digitizing that same song and making it accessible to any visitor to one's Web site would not be. A federal district court in California ruled that the posting of entire newspaper articles to a bulletin board Web site was not a fair use.[9] The court was not persuaded, despite the arguments that the Web site selected only some articles from the newspapers, provided commentary and criticism, and was a not-for-profit service (addressing some of the four fair use guidelines). Imagine, however, the same situation within the walls of a classroom. If a faculty member clipped articles from a newspaper and posted them, in their entirety, on a physical rather than a virtual bulletin board, it seems apparent that there would be no claim by the newspaper of a copyright infringement. It is indisputable that protected uses within a classroom may cease to be protected when they are moved online.

To provide some direction to educators, the Educational Multimedia Fair Use Guidelines Committee provided a set of guidelines on the amount of copying that would constitute a fair use for multimedia. These guidelines were adopted by the House Subcommittee on Courts and Intellectual Property.

The guidelines provided are at present the standards that have been agreed to by representatives of government, education, and content creators. That is not to say that the courts will limit fair use to *only* these uses, nor does it guarantee that anyone adhering to these guidelines will be protected. It is also possible that at some point in the future these numbers may be modified. The standards do provide an agreed-upon limit that *should* be honored by the courts. Faculty ought to be aware of the current standards, make their students aware of the standards, and adhere to them. It ought to be remembered that the above table only addresses the issue of "amount and substantiality" of the work used under the Fair Use Doctrine. The other three fair use requirements must also be met to qualify.

In our digital age, many faculty members are requiring students to use the Internet for class projects. Copyright issues present themselves here as well. If students use a piece of copyrighted music for the music bed in a

Table 9.1
Fair Use Guidelines for Educational Multimedia

Medium	Proportion	Limits	Additional guidelines
Film/Video (Motion Media)	10%	Up to three minutes from a work	
Text	10%	Up to 1,000 words	
Poetry	Entire poem if 250 words or less	Up to 250 words	If from anthology, limit of three from one poet or five poems total
Music/Lyrics/ MusicVideo	10%	30 seconds maximum	No change to melody or fundamental character of the work
Photos/ Illustrations	Complete image	Up to five from one artist	
Collected Work Images	10%	Up to 15 images	Max. 15 from a single collected work
Database (Numerical Data Sets)	10%	Up to 2,500 fields or cell entries such as spreadsheet	Cell entry is row by column intersect (spreadsheet). Field entry is single item of information.

Source: Consortium of College and University Media Centers, 2001.

video production class, the use is likely to qualify as a fair use. Once the work is disseminated beyond the classroom, however, that may not be the case.[10] Faculty members who require their students to post such projects to an online site for downloading or streaming are promoting copyright violations. As strictly a class assignment, student use of copyrighted music in this way could be viewed as analogous to copying a page from a magazine or encyclopedia to illustrate a point made in a term paper. The use of copyrighted material is confined to a rather small audience: the teacher or, if the project is displayed, other members of the class. Broader dissemination of the project online makes the potential audience larger and might affect the impact on the market, which is the fourth criterion to be considered under the Fair Use Doctrine.

Professors who require students to post their papers or projects online ought to be extremely diligent in explaining the limits placed on them by copyright law. It needs to be clear to our students that copyright laws prohibit them from stealing the work of others.

LINKING

The process of providing a direct connection from one Web site to another is known as linking or hyperlinking. Links are added to Web pages at appropriate places to provide an opportunity to be instantly transported to another Web site containing information related to the word or picture that was linked. Links are commonly provided by Web page authors to sites over which they have no control. Are there legal liabilities associated with linking? Is the owner of a Web site liable for misinformation provided by a page to which the original site is linked? Can the owners of Web site prevent others from linking to their sites?

The good news is that there is no legal obligation to obtain permission simply for placing a link on your page to someone else's site. Most consider it proper "netiquette" (net etiquette) to inform the owner or Web manager of the linked site. Since most content creators are seeking the broadest possible dissemination, they do not object to links, provided the links do not disparage them.

Most Web sites consist of dozens of pages, sometimes hundreds, established in a hierarchy, with the home page serving as the "front door." Home pages provide the index or contents for a Web site and serve as the portal through which most web traffic flows. In commercial Web sites, because the home pages get the most hits, they usually contain the preponderance of advertising, whether for the Web site owners themselves or for others who have purchased advertising on that site. In commercial contexts if not all, Web site creators prefer that visitors pass through the homepage first before viewing other pages. When links are created, they can either direct the viewer to a home page or to a more specific page further in the Web site structure, known as "deep linking." Because deep linking can avoid not only advertising but also much of the identifying content of the Web site owner, most Web site owners prefer that visitors be directed to their home pages. Ticketmaster reached a settlement with Microsoft to stop the software giant from deep linking to its site (Kelly and Hieber, 2001). The main complaint by Ticketmaster was that visitors directed to pages by Microsoft's *seattlesidewalk.com* Web site were not even aware that they had been transferred to Ticketmaster. Subsequently, Ticketmaster filed suit against another provider that deep linked to its site but lost the case. Visitors to *Tickets.com* who wanted to purchase tickets sold exclusively by Ticketmaster were linked to the Ticketmaster site. The court found that Ticketmaster pages were clearly labeled as such, and *Tickets.com* even notified visitors that they were being transferred to "another ticketing company."[11] As with many legal issues involving newer technologies, this area of the law is still evolving.

It appears safe to say that deep linking will not create problems for educators unless they somehow misrepresent the material to which they link.

This can sometimes occur, intentionally or unknowingly, when a Web site operator uses frames on screens to any linked sites. "Framing" allows the earlier site visited to provide a frame through which the visitor sees subsequently visited sites. A problem can arise because the frame may obscure content from the subsequent frames, including advertising or content provider identification. In two different cases, one from the East Coast and one from the West Coast, courts have asserted that framing that obscures advertising and logos on framed sites is an infringement.[12] The safest course of action for educators would be to avoid framing when linking.

BULLETIN BOARDS/DISCUSSION GROUPS/CHAT ROOMS

More and more classes are including some form of online discussion as part of their course requirements. For example, a media ethics class might require class participation both in the form of talking in class and in posting to a class discussion group online following the in-class discussion. Are there any legal expectations on a faculty member who allows or requires students to post messages to a publicly available site? While there are no legal cases that deal directly with this issue, there are some cases that provide guidance in this area. As the practice becomes more commonplace, common law on the subject is likely to be developed.

Some faculty members allow students to post messages to class discussions anonymously, believing this will afford students the greatest opportunity to be open and honest. Faculty who choose to operate online class discussions this way need to be especially cautious. Two questions can arise from this anonymous posting of material: Can the operators of such bulletin boards be held legally responsible for all materials posted to their sites, and can the operators be made to disclose the identities of posters who defame others? While there are no cases dealing specifically with educational bulletin boards, there are dozens of cases involving similar venues.

There have been conflicting court decisions regarding the legal liability of individuals who operate bulletin boards. In a well-known case involving CompuServe, a federal court compared such services to bookstores and libraries, holding that they could not possibly be responsible for all the content within them.[13] The New York Supreme Court held that Prodigy was different from CompuServe in that Prodigy acted more like a publisher in deciding the content of postings. Prodigy made decisions about what sorts of messages were acceptable and unacceptable and attempted to prevent unacceptable messages from being posted. As such, the court held that Prodigy was responsible for content posted on its bulletin boards because it exercised a form of editorial discretion.[14] Many have differentiated between the cases because of Prodigy's active involvement in filtering out unacceptable messages as compared to CompuServe's more open policy (Herbeck, 1999). That may have been true in the 1990s when the cases were decided, but that distinction may no longer be valid as cyber law continues to evolve. The New Jersey Supreme Court held that an employer could be held responsible for sexual harassment that occurred on a work-related bulletin board. The court stated that an employer that provides a forum for harassment and does nothing to stop it sends a message that such harassment is acceptable ("You May be Liable . . .," 2000). While the case dealt with employers and their relationship to employees, its comparison to faculty and their relationship to students is easily made. If the only distinction between CompuServe and Prodigy is the amount of "editorial discretion" exercised, the New Jersey court creates a Catch-22 by holding a bulletin board liable for not exercising editorial discretion but making it liable to other legal challenges if it does. One other case involving electronic bulletin boards, this time with regard to copyright, provides one other bit of guidance. A federal court stated that while bulletin board operators cannot possibly know everything posted on their sites, they are obligated to take action when they have been notified of possible unlawful activity. When Church of Scientology materials were posted to a bulletin board, the bulletin board operator could not have known there was a problem. Once notified by the church, the operator could have acted more responsibly.[15] Clearly we have not heard the last word on the liability of bulletin board operators.

One of the hottest topics currently online deals with "cybersmear," the process of maligning a person or company online using publicly accessible discussion groups. These attacks are made anonymously and, in cases where the authors have been identified, have come from disgruntled employees or dissatisfied customers. In several cases, the defamed have successfully convinced courts to require the bulletin board operator to divulge the identity of the persons responsible for the actionable comments (Sobel, 2000). While critics complain that companies force disclosure merely to chill speech that they don't like, the reality is that these suits are likely to continue unless legislation arises to prevent them, which does not appear to be in the offing. At the very least, faculty who provide anonymous bulletin boards or discussion groups for their classes ought to strongly advise students that no postings should be defamatory or disparaging, and that doing so could put them at risk of having their identities revealed.

One other prudent precaution to take is to password-protect any class discussions or bulletin boards, enabling only class members to access the site. While students might still be liable for any defamation that occurs in the discussion, the damages will clearly be limited by limiting the scope of who may receive the communication. If only the 30 members of a class, or even 300, are able to see a potentially damaging remark, the extent of the damage is likely to be less than if thousands see the remarks. From a pedagogical perspective, this is also appropriate, as most faculty only want participation from students who have the necessary content from class readings and lectures, rather than outsiders who come to the discussion unprepared.

PRIVACY

College students enjoy the same rights of privacy that are protected for other adults in American society. In some regards their rights to privacy are even more specifically enunciated than they are for the general population. The Family Educational Rights and Privacy Act (usually referred to as the Buckley Amendment) protects a student's right to privacy on campus. First enacted in 1974 to protect the confidentiality of student educational records, the law has been amended several times. Among other things, it is the set of rules that prohibits faculty members from disclosing the grades of any student except under specific, enumerated circumstances. Despite ignorance by many college faculty, posting of student grades by any easily identifiable means (including Social Security numbers or in alphabetical order) is a violation of the Buckley Amendment.

Generally speaking, the only information that educational institutions ought to be making available about their students is what has been termed "directory information," which includes

> student's name, address, telephone listing, electronic mail address, photograph, date and place of birth, major field of study, dates of attendance, grade level, enrollment status (e.g., undergraduate or graduate; full-time or part-time), participation in officially recognized activities and sports, weight and height of members of athletic teams, degrees, honors and awards received, and the most recent educational agency or institution attended.[16]

It is important to note that these rules prohibit disclosure by the institution, or the requirement by faculty that students disclose personal information. It does not preclude a student from volunteering to disclose any personal information on a Web site of his or her creation.

If, for example, a faculty member requires that students in a class create personal Web sites, the faculty member could list items that should be incorporated into the sites, including those listed above. If, however, the requirement includes that the students list their grade point averages, marital status, health conditions, or the names of family members, students could claim their privacy rights would be violated by such a requirement. On the other hand, if students are allowed to choose what to include on their Web sites and elect to self-disclose some of the information, there is no legal violation.

OTHER APPLICABLE LAWS

The development of the Internet did not create copyright violations, but it did make it possible for them to occur more frequently, and likewise to be detected more efficiently. Other rules that have always been applicable to educators continue to be applied in the online world. Two examples of such rules are the enforcement of laws prohibiting trademark infringement and product disparagement. While these are rarely problems for educational institutions, they can become problems in an Internet world.

Trademark infringement usually occurs when someone uses a name or trademark belonging to another company in such a way as to confuse a consumer. Clear examples of this occur when someone tries to market a product using another company's logo, such as putting Nike or a trademarked "swoosh" on a piece of athletic apparel. Cases that are not as clear might involve a company trying to sell T-shirts with a swoosh that was similar to, but not quite the same as, the famous Nike trademark. A determination of whether a consumer is likely to be confused is a subjective judgment, and there are degrees of similarity that are considered. Trademark infringement is most often a conflict between two commercial entities, which is why educators and educational institutions are rarely involved.

Where educators can run afoul of trademark law is if they use a trademark or logo in their Web site or multimedia productions that implies an endorsement or affiliation with a particular product. If a professor creates a CD-ROM to teach video production, and in the production the subject drinks from a can of Coca-Cola, there is no trademark infringement. On the other hard, if there is any statement that implies that the CD is "brought to you by" Coca-Cola, or sponsored by Coke, then Coca-Cola had better actually have sponsored the production. Merely using a product is not an infringement, whether on a Web page, on a CD, or even in a television program on a commercial network.

Product disparagement occurs when, as the words suggest, a product is criticized in a manner to cause damage to its reputation. In the above example, no product disparagement occurs in the CD-ROM if a subject is seen drinking a Coca-Cola. Product disparagement would occur if the subject who drank it then began gagging and fell to the floor screaming.

SIMPLY PUT . . .

Educators need to be aware that they are not immune from copyright, libel, privacy or other legal actions simply because they are fulfilling an educational mission.

NOTES

1. Marobie-Fl v National Association of Fire Equipment Distributors, 983 F. Supp. 1167 (E.D. Ill. 1997).
2. Encyclopaedia Britannica Educ. Corp. v Crooks, 558 F. Supp. 1247 (W.D.N.Y. 1983).
3. Marcus v Rowley, 695 F. 2d 1171 (9th Cir. 1983).
4. 17 USC §107.
5. Copyright Office, *Reproduction of Copyrighted Works for Educators and Librarians*, p. 8.
6. Harper & Row Publishers v Nation Enterprises, 471 U.S. 539 (1985).
7. Basic Books, Inc. v Kinko's Graphics Corp., 758 F. Supp. 1522 (S.D.N.Y. 1991) and Princeton University Press v Michigan Document Service, 99 F.3d 1381 (6th Cir. 1996), cert. Denied, 520 U.S. 1156 (1997). The Michigan case was earlier decided in favor of the copy center, but the decision was reversed in favor of the original copyright holders.
8. Faculty might also want to consider the possibility of password protecting their sites to limit access to students enrolled in their classes. While this is counter-productive to those seeking broad dissemination, it does afford greater control over an author's works.
9. Los Angeles Times v Free Republic, 54 U.S.P.Q.2d (BNA) 1453 (C.D. Cal. 200).
10. At least one federal court has found, however, that use of music as background for an educational video may be a fair use, provided the amount is still small. A federal court in Michigan ruled that a public television station did not violate copyright by using a 30-second segment of a popular song as background for opening scenes. Higgins v Detroit Educational Broadcasting Foundation, 4 F. Supp. 2d 701 (E.D. Mich. 1998). Faculty ought to be very cautious, however, in expecting this decision to apply outside of Michigan.
11. Ticketmaster v Tickets.com, CV 99-7654 (C.D. Cal. March 27, 2000).
12. Futuredontics v Applied Anagramic, 45 U.S.P.Q. 2d 2005 (C.D. Cal. 1998) and Washington Post v Total News, No. 97-1190 (S.D.N.Y. Feb. 20, 1997), settlement reached June 5, 1997.
13. Cubby v CompuServe, 776 F.Supp. 135 (S.D.N.Y. 1991).
14. Stratton Oakmont v Prodigy, 23 Media Law Reporter 1794 (1994).
15. Religious Technology Center v Netcom, 907 F.Supp. 1361 (N.D. Cal. 1995).
16. 20 U.S.C. 1232g (a) (5) (A).

REFERENCES

Braun, Mark, 1997. "Broadcast Educators' Knowledge of Provisions of U.S. Copyright Law and the Doctrine of Fair Use: New Technology Implications." Paper presented at the Broadcast Education Association Convention, Las Vegas, April.

Consortium of College and University Media Centers, 2001. "Fair Use Guidelines for Educational Multimedia." http://www.indiana.edu/~ccumc/guidesum.html

Copyright Office, 1998. *Report on Copyright and Digital Distance Education.* Washington, DC: Library of Congress.

Copyright Office, 1999. *Reproduction of Copyrighted Works for Educators and Librarians.* Washington, DC: Library of Congress.

Crews, Kenneth, 2000. *Copyright Essentials for Librarians and Educators.* Chicago: American Library Association.

Fishman, Stephen, 2000. *The Copyright Handbook: How to Protect and Use Written Works*, 5th ed. Berkeley, CA: Nolo.

Goldstein, Paul, 1999. *Copyright, Patent, Trademark and Related State Doctrines, Cases and Materials on the Law of Intellectual Property*, 4th ed. New York: Foundation Press.

Halpern, Sheldon, C. A. Nard and K. L. Port, 1999. *Fundamentals of United States Intellectual Property Law: Copyright, Patent and Trademark.* London: Kluwer.

Herbeck, Dale, 1999. "Cubby or Stratton Oakmont? Defamatory Speech on Computer Bulletin Boards," in *Real Law @ Virtual Space*, Susan Drucker and Gary Gumpert, eds. Creskill, NJ: Hampton Press.

Jasper, Margaret, 2000. *The Law of Copyright*, 2nd ed. Dobbs Ferry, NY: Oceana.

Kelly, David and Christina Hieber, 2001. "Confusion, Dilution Still the Focus of Linking Cases." *National Law Journal*, 5 Feb.

Lehman, Bruce, 1998. *The Conference on Fair Use: Final Report to the Commissioner on the Conclusion of the Conference on Fair Use.* Washington, DC: U. S. Patent and Trademark Office.

Litman, Jessica, 2001. *Digital Copyright: Protecting Intellectual Property on the Internet.* Amherst, NY: Prometheus.

Sappenfield, Mark. 2001. "Internet as Portal to Ivory Tower." *Christian Science Monitor*, April 6, p. 1.

Shaw, Paul, 1998. *Managing Legal and Security Risks in Computing and Communications.* Boston: Butterworth-Heinemann.

Sobel, David. 2000. "The Process That 'John Doe' is Due: Addressing the Legal Challenge to Internet Anonymity," Vol. 5. *Virginia Journal of Law & Technology*, p. 3.

Stim, Richard, 2000. *Getting Permission: How to License and Clear Copyrighted Materials Online and Off.* Berkeley, CA: Nolo.

"You May Be Liable for Employees' On-Line Harassment," 2000. *Maine Employment Letter*, Sept.

CHAPTER 10

Examining Race and Gender in Student Use and Evaluation of New Technologies

Mark J. Braun and Rebecca Ann Lind[1]

New technologies are bringing changes to academe, and media education is no exception. New technologies allow students to create and distribute media content in new ways, to participate in asynchronous and distributed group work, to query their instructors at any time, to hand in assignments electronically and to take exams online. In addition, instructors can utilize course management software, e-mail, and listservs; they can post grades, extend class discussions, and distribute assignments and syllabi via the Web; they can encourage knowledge application, relate course content to actual practice in the professional world, and better serve as mentors or coaches by providing individualized and timely feedback (Groves & Zemel, 2000).

This chapter considers race/ethnicity and gender issues in students' use of new technologies. It reports the results of a survey investigating which classroom technologies students have used, how much, when, why, with what comfort level, and with what level of facility.

This research has implications for educators striving to improve pedagogy via new technologies. We provide a baseline figure of technology usage in 2001 against which to compare future use levels. More importantly, we investigate to what extent race and gender are related to the adoption of, and level of comfort with, new technologies. If certain groups are uncomfortable with or avoid using new technologies, members of those groups may find their learning efforts hampered. As Cookman (1998, p. 37) stated, these "are not ordinary tools such as a hammer and saw but sophisticated vehicles for more powerful learning. Students will have to be equipped with the necessary computer skills as a minimal requirement to avoid frustration on their part."

However, our investigation of racial and gender issues in the use and evaluation of new technologies goes beyond comparing skill levels. We do not presume differences in aptitude. Rather, it is the social construction of technology, and the interaction of these socially constructed technologies with social constructions of race or gender, that intrigues scholars (Brunner & Bennett, 1998; Harrell, 1998).

LITERATURE ON COMPUTERS AND GENDER

Research investigating the roots of gender differences has indicated that school-age boys use home computers more than do girls (Robertson et al., 1995), especially computer games (Culley, 1993). Male students receive more encouragement regarding computer use from friends than do female (Busch, 1995), and in mixed-sex school groups tend to dominate computer-based activities (Underwood & Geoffrey, 1990). Overall, girls are likely to spend less time on computers than are boys (Kayany and Yelsma, 2000; WEEA Digest, 2001). Until about fifth grade, girls and boys appear equally interested in computers, but, probably because of socialization differences, boys' use rises significantly in junior and senior high, while girls' drops (Barbieri & Light, 1992; Kantrowitz, 1994).

Gender-socialization carries over into the college years. College students demonstrated significant gender differences in computer self-efficacy (Miller, 1996). Schumacher and Morahan-Martin (2001) found that female students reported greater discomfort and incompetence with their use of computers, while males reported more experience with computers, programming, and graphics/games applications. Gender differences were also found in online media proficiency (Kayany & Yelsma, 2000) and anxiety and affective states related to computer use (Coffin & MacIntyre, 1999; McDowell & Schuelke, 1998). As did younger girls, female college students valued computer technology in a social context, while males viewed computers as objects. Men described "an early and persistent magnetic attraction" to computers, while women linked computer science to "a larger societal framework," attaching their interest to specific contexts, such as education, medicine, space exploration, or the arts (Margolis, Fisher, & Miller, 1999/2000, p. 13).

Males and females also differed in their reports of incentives, barriers to technology use, and factors related to use. Specifically, females rated time to learn a technology, ease of use, training, increased student learning, and availability of disciplinary information as more important than did males (Spotts, Bowman, & Mertz, 1997), although some differences may be field-specific (Huppert, Yaakobi, & Lazarowitz, 1998).

Gender (and ethnicity) were not seen as significant factors in the degree of utilization or level of satisfaction with student services provided to students in distance-education courses (Paneitz, 1997). However, communication patterns in distance education courses revealed gender differences, with women emphasizing intimacy, symmetry based on the establishment of similar experiences, status equality, and interdependency (Gougeon,

1998). Reinforcing the importance of social organization, Harrell (1998) found that groups based on cooperative principles fostered positive attitude formation and more equal participation by men and women.

LITERATURE ON COMPUTERS AND RACE / ETHNICITY

The literature in this area suggests that while technology has greatly changed how students learn and communicate, a lack of parity may leave some minority students disadvantaged. Again, we don't assume skills-based differences; motivation- or habit-based explanations (Pingree et al., 1996, p. 334) seem much more fruitful. Access is also an issue. Black, Hispanic, and Native American children have as much as 30% less access to the Internet than White children (Malveaux, 1997), and Asian Americans have far greater access than do Whites (U.S. Department of Commerce, 2000). African American households with incomes under $40,000 are significantly less likely to have a computer at home and use the Internet—only half as likely as similar White U.S. households (Novak & Hoffman, 1998; Raloff, 1998). Further, White students without a home computer are much more likely to find an alternate source of access than are Blacks (U.S. Department of Commerce, 2000).

The racial computer gap appears to carry over to schools as well; urban districts with more Black students supplied significantly fewer computer instruction opportunities than did other urban districts (Hess & Leal, 1999). Native American students living on reservations may be unable to access the Internet efficiently. Some reservation schools have installed sophisticated hardware to support high-quality Internet access, but they can't get anyone to connect a fast broadband line in a relatively remote area (Zehr, 2001).

This lack of parity could carry over into students' college years in terms of comfort with the use of technology. Hawkins and Paris (1997) found White undergraduates more proficient in information technology skills, and that Black students preferred Macintosh computers, while Whites preferred PCs. Minority students also spend more time with computer-assisted instruction programs (Besser, 1993).

Also, social differences in technology use are evident. For instance, African Americans use e-mail more often than do Whites for both job-related and educational purposes, while Whites use the Internet more for shopping and bill paying than do Blacks. Whites are more likely to use search engines, seek non-commercial information online, and search the Web for product and company information prior to making a purchase (U.S. Department of Commerce, 2000).

Pingree et al. (1996, p. 333) suggested that because "African-Americans and Hispanics are on average poorer and have less formal education than Whites, new technologies run the further risk of being culturally insensitive or being seen as 'White' and thus irrelevant." However, computer-mediated classrooms may also be great equalizers, empowering minority students by providing easier access to "the floor" during electronic discussions (Lewis, Whitaker, & Julian, 1995).

INVESTIGATING STUDENT USE AND EVALUATION OF TECHNOLOGIES

During spring 2001, we asked 900 educators to invite their students to participate in an anonymous Web-based survey. The sampling frame was drawn from e-mail addresses in the Broadcast Education Association and National Communication Association membership directories and lists of historically Black colleges and universities, Hispanic-serving institutions, and Native American tribal colleges.

The survey focused on classroom technologies (e-mail, bulletin board, discussion list, group conferencing, distance learning, and online quizzes, tests, and homework). Students were asked how *often* they used the technologies, how *comfortable* they are with each, how *easy* each is to use, how *helpful* each has been when used in connection with college courses, how much *training* they received, and how much *support* is available. Students also indicated which technologies had been most and least beneficial in connection with a college course.

Open-ended questions asked why, what makes the technology easy and hard to use, and what instructors could do to make it easier to use. Finally, students provided demographic information.

A total of 237 surveys was completed. The sample was 46.9% male and 53.1% female; 76.4% Euro-American, 2.2% African American, 5.3% Latino/a, 9.3% Asian/Pacific Islander, 0.9% Native American, and 5.8% bi/multi racial. Only 13.2% were freshmen or sophomores, with 40.4% juniors, 39.9% seniors, and 6.6% graduate students. Less than 10% reported majoring in anything other than a variant of communication/mass communication/journalism.

Most of the sample accessed technologies from home (88.2%) or school (81.2%). Far fewer did so from libraries (56.5%), work (26.2%), or cybercafes (3.0%). We found that 37.9% used at least one of the standard office applications (word processing, spreadsheets, presentation software) but no other type of application such as Web authoring or graphics/video software. Many respondents (44.7%) used at least one office-type application as well as *either* Web authoring applications or video/graphics applications, while 17.4% used all three types of applications.

EXTENT OF TECHNOLOGY USE

Students indicated how much they used each technology. Table 10.1 shows that e-mail is the only technology that is nearly universally "often"

Table 10.1
Student Usage and Evaluation of Technologies

	E-mail		Bulletin Board		Discussion List		Group Conferencing		Distance Learning		Online Quizzes	
	Mean	SD	Mean	SD	Mean	SD	Mean	SD	Mean	SD	Mean	SD
How Often Used	2.71	0.55	0.94	0.98	0.85	0.91	0.57	0.82	0.46	0.78	1.57	1.00
3 = often, 2 = sometimes, 1 = rarely, 0 = never												
Comfort Levels	4.87	0.42	3.92	1.01	3.64	1.14	3.62	1.00	3.49	1.08	*	*
5 = very comfortable, 4 = moderately comf., 3 = neutral, 2 = mod. uncomf., 1 = very uncomf. **Due to technical difficulties with the survey, data from this variable cannot be analyzed.*												
Ease of use	4.95	0.28	4.31	0.77	4.06	0.95	3.88	0.91	3.91	1.00	4.40	0.74
5 = very easy, 4 = moderately easy, 3 = neutral, 2 = mod. hard, 1 = very hard												
Helpfulness	4.82	0.47	3.82	0.88	3.57	0.96	3.43	1.01	3.59	1.04	4.09	0.87
5 = very helpful, 4 = mod. helpful, 3 = neutral, 2 = slightly helpful, 1 = not at all helpful, 0 = actually hurt rather than helped												
Training Received	1.91	0.94	1.85	0.83	1.82	0.84	1.82	0.88	1.82	0.84	2.13	0.85
4 = too much training, 3 = just enough training, 2 = very little training, 1 = no training at all												
Support available	3.24	0.64	2.76	0.74	2.61	0.73	2.63	0.83	2.72	0.87	2.91	0.68
4 = more support than needed, 3 = just enough support, 2 = very little support, 1 = no support at all												

(continues)

Table 10.1 (*continued*)
Student Usage and Evaluation of Technologies

	E-mail		Bulletin Board		Discussion List		Group Conferencing		Distance Learning		Online Quizzes	
	Mean	**SD**	**Mean**	**SD**	**Mean**	**SD**	**Mean**	**SD**	**Mean**	**SD**	**Mean**	**SD**
	Freq	**%**	**Freq**	**%**	**Freq**	**%**	**Freq**	**%**	**Freq**	**%**	**Freq**	**%**
Most Beneficial	194	85.5	11	4.9	1	0.4	3	1.3	4	1.8	13	5.8
Least Beneficial	3	1.5	40	19.9	26	12.9	55	27.4	48	23.9	29	14.4

Full breakdowns by race and gender are available at www.uic.edu/~rebecca/newtechreport.html

used. Except for online quizzes (used less than "sometimes" but more than "rarely"), the remaining technologies are rarely used. Distance learning is used the least.

To investigate possible race- or gender-based differences in the extent of technology use, a series of T-tests (gender) and one-way ANOVAs (race/ethnicity) considered the amount of technology use as dependent variables. Because of the small numbers of minority respondents, but being mindful of prior research findings, we recoded race into "White" (n = 172), "Asian" (n = 21), and "Other" (n = 32). Respondents who never used a given technology were excluded from the analysis.

Of the 12 tests (6 *t*-tests and 6 ANOVAs), none showed any significant race- or gender-based difference in the extent of any technology usage.

COMFORT WITH TECHNOLOGY

Table 10.1 presents the means and standard deviations of responses to questions asking how comfortable students were with each technology. Our respondents were relatively comfortable with all of the technologies; no mean was any lower than the midpoint between "neutral" and "moderately comfortable," although large standard deviations indicate great variation among responses. Students were most comfortable with e-mail—the mean is close to the maximum "very comfortable" and the standard deviation much smaller than with any other technology.

Five *t*-tests comparing men and women on comfort with each technology revealed no significant differences. Of five ANOVAs comparing racial/ethnic groups, two (group conferencing, distance learning) generated significant differences, but only one remained when the follow-up Scheffe test was conducted (unsurprising, given the large variances). Whites seemed more comfortable than Asian Americans in using group conferencing technologies. We minimize the significance of this finding because over all the statistical tests we conducted, only about 5%—what we'd expect by chance—found significant results. As such, we will not follow conventions of providing test statistics values, degrees of freedom, and significance level.

EASE OF TECHNOLOGY USE

Table 10.1 indicates that all the technologies seem fairly easy to use, with only group conferencing and distance learning falling (slightly) below the "moderately easy" level. E-mail is by far the easiest technology for respondents, with online quizzes coming in second.

Six *t*-tests comparing each technology's ease of use across men and women generated one significant difference: females rated online quizzes easier to use than males did. Of six ANOVAs testing for racial/ethnic differences, two (group conferencing, distance learning) were significant. Again, only one remained after the Scheffe test: Whites rated distance learning easier to use than did Asian Americans.

TECHNOLOGY'S HELPFULNESS

As Table 10.1 reveals, the six technologies grouped into three tiers of perceived helpfulness. E-mail was rated most helpful, almost achieving the maximum "very helpful" value. The middle tier (online quizzes, bulletin boards) clustered around "moderately helpful." Group conferencing and distance learning created the bottom, least helpful tier.

None of the six *t*-tests for gender differences in helpfulness of technology or six ANOVAs for race differences generated significant results.

TECHNOLOGY TRAINING

Table 10.1 shows that most students do not believe they're getting enough training for any of the technologies. No technology even comes close to having "just enough" training. Only online quizzes was rated (slightly) above the "very little" level. Students report receiving even less than "very little" training in the use of all other technologies.

Six *t*-tests (gender) and six ANOVAs (race) compared groups' perceptions of the adequacy/extent of training they'd received in using each technology. None provided significant results.

TECHNOLOGY SUPPORT

As Table 10.1 reveals, e-mail is the only technology that is perceived to be supported sufficiently. Online quizzes receive almost enough support, with the remaining technologies hovering between "very little" and "just enough" support.

Of the six *t*-tests (gender) and six ANOVAs (race) comparing groups' perceptions of the adequacy/extent of technological support, none was significant.

MOST AND LEAST BENEFICIAL TECHNOLOGIES

After evaluating each technology individually, respondents were asked to select the one most beneficial and one least beneficial technology. E-mail was deemed most beneficial by most students; it was selected by 85.5% of respondents. Table 10.1 shows that a total of only 32 students chose anything other than e-mail as most beneficial, with the bulk of those fairly evenly divided between bulletin boards and online quizzes. Two chi-square analyses tested for gender- and race-based differences in most beneficial technology, but neither was significant.

Far more variation exists among the least beneficial technologies. As Table 10.1 shows, discussion lists and online quizzes each were selected least beneficial by more than 10% of students, while nearly 20% chose bulletin boards. Two technologies were dubbed least beneficial by approximately one-quarter of respondents: group conferencing and distance learning.

Two chi-square analyses compared gender- and race-based groups' selection of least beneficial technology. Neither revealed any significant differences.

BEHIND THE NUMBERS: ANALYSIS OF OPEN-ENDED RESPONSES

After indicating which technologies were most and least beneficial, respondents were asked open-ended questions about each. Each question was analyzed separately, and responses were closely read to determine the main themes they contained. These themes were defined, and responses were coded for the presence of each theme. Themes mentioned by at least five respondents are presented below in the rank order of their use (based on frequency of mention across the sample).[2]

After coding open-ended responses for each theme's presence, we tested for significant race- or gender-based differences in the use of each. This might reveal differences obscured by the use of quantitative survey data: Do women perceive more technological limitations than men do, or value technologies for assisting in interpersonal contact more than men do, or say that technologies lack the "human touch" more often than men do?

Testing for race and gender differences in the use of each of the 46 themes presented below required a total of 92 separate chi-square analyses. Only four revealed significant differences, three of which were based on expected cell counts lower than the minimum of five generally required by the statistic. Thus, the only difference that can be reliably interpreted is that Whites are far more likely than Asians to say that a technology is deemed "not beneficial" because of technological drawbacks. However, given the overall pattern of nonsignificance across this large number of statistical tests, we caution against too much stock being placed in this finding. As with the quantitative survey data, we have found no pattern of gender or racial/ethnic differences in the use or evaluation of new technologies.

What Makes Beneficial Technologies Beneficial?

Nine themes appeared in the responses to this question asking what makes the most beneficial technology beneficial.

1. The dominant theme, appearing in 52.6% of responses, is *Personal Contact with Professor*. This reflects students' use of technology to communicate one-on-one

with their teachers, not merely to receive information sent to the entire class. Examples include: "technology such as e-mail allows students to be in close and constant connection with teachers," and "facilitates frequent contacts between teachers and students."

2. The second-ranked theme is *Contact with other Students*. This theme, in 26.3% of responses, reflects technologies fostering interpersonal communication among students. Examples include: "it allows students to easily converse while not in the classroom setting" and "students can stay in touch very easily."
3. The third-ranked "beneficial" theme (25.4% of responses) addresses *Ease; Convenience.* This reflects students' perceptions that a technology is easy or convenient to access or use. Examples include: "access for students is easy," and "access to e-mail is very abundant."
4. The fourth-ranked theme is *Time Flexibility*. Mentioned by 23.2% of respondents, this reflects using technology to conduct academic business whenever students choose. Examples include: "it allows me to take a test on my own time," and "I am able to respond, print out, etc., at my own leisure."
5. In the fifth "beneficial" position is *Course Support*, seen in 21.5% of responses. This theme reflects using technology to support the learning objectives of a course—providing course material, helping students learn from each other, and reinforcing class content: "if you miss a day of class the assignments are on the Web," "you can share ideas with other students," and "[it] expands on the learning process [and] allows for learning outside of the classroom."
6. The sixth-ranked theme, in 20.6% of responses, is *Coursework Management*. This reflects using technology to submit assignments or track one's course progress: "it allows me to turn in papers and assignments later than the class meetings," and "having written record of assignments (incoming and outgoing) is also a benefit."
7. In 18.0% of responses, the *Efficient; Saves Time* theme reflects students' use of technology to complete tasks and conduct academic business as quickly as possible: "you can communicate with your teachers and other students without having to spend a lot of time playing phone tag," and "it can be more efficient than calling or going in person."
8. The eighth-ranked theme, in 3.9% of responses, is *Reduce Anxiety; Increase Confidence.* These responses indicate that technology can help students overcome personal issues of nervousness and general discomfort: "allows me to become familiar with tests/quizzes/etc. without any test anxiety," and "it helped me to be comfortable in front of others that weren't in the same classroom as me."
9. Ranked ninth, mentioned by only 2.6% of respondents, is *Saves Paper*. This theme reflects students' desire to conserve paper: "it also allows students to turn in assignments and things like that to the professor in a way that doesn't require wasting paper" and "[much better] than handing out reams of paper."

What Makes Beneficial Technologies Easy To Use?

When students were asked what makes the technology they'd selected as "most beneficial" easy to use, seven themes appeared.

1. The most common theme, appearing in 35.5% of responses, is *Simplicity*. Examples include: "it's not rocket science. Even my 5-year-old nephew knows how to surf the Web and use e-mail. Anybody can use it as long as they are not illiterate or brain dead" and "it is fairly straightforward."
2. Ranked second, in 27.2% of responses, is *Convenience*. This theme refers to the convenience and flexibility afforded to individual users: "it allows me to work on my own time, when I want" and "convenient to use and have."
3. Used in 19.7% of responses, the theme *Used widely* indicates that beneficial technologies are easy to use because they are ubiquitous: "because everyone has an e-mail account, so it's easy for everyone to access," and "e-mail is an ubiquitous part of the university community."
4. The theme ranked fourth (9.6% of responses) is *Experience*. Ease here stems from length or history of use. Examples include: "it's easier because I've been using it longer than any of the other technologies," and "I've been using e-mail since the early 1990s and I am very comfortable with the e-mail process."
5. Ranked fifth, in 7.9% of responses, is *Used often*. This theme refers not to longevity but rather frequency of personal use: "it is something I use with my friends quite often and have become good at" and "it's in use every day."
6. The sixth-ranked theme was evident in 5.3% of responses: *Course Support*. This refers to technologies being easy because they are good sources for course-related information: "the classes that I take with online versions of the textbook are of great value in helping get the exact information that I need to learn."
7. Found in only 2.6% of responses was the *Training* theme. This reflects students' indications that they find, or would find, the technology easy to use because of more training in the technology. Examples include: "they can be overwhelming the first time you attempt to tackle something" and "as much as I hate the thought of taking more requirements, . . . a basic computer class should be required."

What Makes the Most Beneficial Technologies Hard to Use?

When students were asked to consider the downside of the "most beneficial" technology and think about what makes it hard to use, five themes emerged.

1. The most common theme, appearing in 28.1% of responses, is *Technological Limitations*. The technology may be slow, difficult to use, crash, or allocate insufficient resources for storage; there may be platform or application difficulties, or there may be fears of delivery failures. Examples include: "if someone sends a file that you do not have the program to open it up in" and "information is lost and not received, which hurts your academic performance."
2. Coming in second is *Nothing*. This theme, in 19.7% of responses, indicates that teachers need do nothing to make the technology easier. The presumption is that the technology is sufficiently easy to use: "I honestly can't think of any aspect that makes this technology difficult to use" and "nothing is hard about it at all for me."

3. Ranked third is *Inaccessibility*, seen in 13.2% of responses. A technology is not always as readily available as it could be: "computer labs are always full" and "not having a computer."
4. The fourth-ranked theme, in 11.4% of responses, is *Insufficient Training*. Students say they need additional training—they don't understand a technology, or aren't comfortable with all of its features. Examples include: "I've never been taught to use some of these technologies," and "if you don't know how to use the technology, it's hard to find someone who can explain it to you."
5. Ranked fifth, in 10.1% of responses, is *Personal Usage Issues*. This theme includes problems resulting from one's own errors, as well as other people's usage habits. For example: "my own error in typing in the addresses," "if people wait to get back to you," and "some teachers don't check their e-mail."

How to Make the Most Beneficial Technology Easier to Use

When asked what instructors could do to make the most beneficial technology easier to use, student responses reflected six themes.

1. The top-ranked theme is *Nothing / Not teachers' job.* Appearing in 44.3% of responses, this theme includes reports that things are just fine as they are, or that it's not the teacher's job to make the technology easier to use. Examples include: "absolutely nothing—couldn't have asked for anything more" and "my instructors do not have the responsibility of making this technology easier—that is why we have a student computing center where we can receive help when needed."
2. Mentioned by 13.6% of students is *Organized content delivery.* This theme includes: "the only thing that should be done is better organization of material."
3. Ranked third is *Training for student*, seen in 11.0% of responses. Examples include: "the instructor should ask if everyone is familiar with that part of technology, and if not, they should see them on how to use it" and "explain what can actually be sent through e-mail to a certain computer. And differences between Macs and PCs when it comes to e-mailing."
4. In 7.5% of responses, the theme *Encourage Use; Use in Class* reflects student perceptions that instructors should use the technology in class. Examples include: "a lot of my profs have really been good about bringing in a laptop, hooking it up to the projector, and walking the class through online assignments, etc." and "more classes should use regular online quizzes."
5. Seen in 5.3% of responses, the *Access* theme reflects students' desires for instructors to provide access to labs, applications, class e-mail rosters, and the teachers themselves: "the teacher can inform the students on when and where open labs that have Internet access are" and "letting his/her e-mail address [be] very accessible."
6. *Training for teacher* is the lowest-ranked theme, seen in 3.5% of responses. This theme indicates that technology would be easier to use if the teachers knew how to use it better: "become more familiar with online tests and lecture/class

notes," "have good Web designing skills," and "be knowledgeable about how e-mail works in case a student has a technical question."

Why Are the Least Beneficial Technologies Not Beneficial?

Students were asked to describe why the technology they'd selected as the least beneficial was not beneficial. Six themes emerged.

1. The most common theme, in 11.8% of responses, was that technology *Lacks Personal Contact*. Technology cannot replace interpersonal communication: "a university campus is a place for people to come together face-to-face to exchange ideas and learn about communicating peer-to-peer. Group conferencing replaces a vital part of the experience" and "it's just easier to meet and talk in person."
2. The second-ranked theme, in 11.4% of responses, is *Inappropriate Use*: a technology is not useful if not used appropriately. Proper use may mean that people should use the technology appropriately, or it may mean that the technology requires multiple simultaneous users who don't happen to be there at the time a student logs on. Examples include: "it requires that others participate, not just you," and "I was required to read and post to the bulletin board, and it was being used as a chat forum, when it should only be used as a posting for events and information."
3. The theme *Inconvenient; Time-Consuming* was seen in 8.3% of responses. The technology takes too much time to use, is tedious, requires too much effort, or is inconvenient: "it is a waste of time" and "it was too big of a hassle to keep going back and reposting arguments to people's comments every few days."
4. The theme *Technological Drawbacks* was found in 7.5% of responses. The technology may be slow or difficult to use, or students may doubt the technology will perform correctly: "it takes too long to get a response, and it's a pain to use the program," "systems go down too often," and "I am leery of it. . . . What if for some reason your quiz or test does not go through the system? There is no way to prove you took it."
5. *Ineffective Tool; Doesn't Add Value*, in 7.0% of responses, ranked fifth. This theme reveals that technology doesn't always facilitate learning, may not provide a worthwhile supplement to classroom activities, doesn't add value, or may not be needed: "you already have a classroom to discuss those topics in," and "everyone looked at it as busywork. I gained nothing."
6. The lowest-ranked theme (3.9% of students) is *Insufficient Training*: "[when I tried it, things] just turned into a bigger mess, maybe lack of training," and "I need more [advice] in using it and it wasn't used often enough to practice."

What Makes the Least Beneficial Technology Hard to Use?

Students were asked to describe what makes the technology selected as least beneficial hard to use. Seven themes emerged.

1. The top-ranked theme, in 13.6% of responses, is *Insufficient Training.* This reflects the need for proper training before a technology can be useful: "I have never been taught to use it," "lack of training," "it's not explained to students," "no one seems to teach how to use it," and "I need more advice in using it and it wasn't used often enough to practice."
2. The second-ranked theme is *Technological Limitations.* In 9.2% of responses, this reflects technological and infrastructure problems: "servers slow down when the load is high, especially if the class is large, and browsers crash in the middle of the project" and "it is just a very clumsy media requiring too many instruments to be working at their peak when it is hardly the case that there is system without any glitches running."
3. Ranked third, in 7.9% of responses, is *Inaccessibility.* This theme refers to the technology's inconvenience: "it's not really hard to use unless someone doesn't have a computer at home" and "it is a pain to log on and do assignments on the computer."
4. Some respondents indicated that just because a technology is not beneficial doesn't means it's hard. The *Not Hard* theme, in 7.5% of responses, includes: "nothing makes it hard to use" and "it's not hard to use, I just haven't had to use it in the past much."
5. The theme ranked fifth, in 6.1% of responses, is *Lack of Personal Contact.* Examples include: "no direct or one-on-one communication with teachers and students" and "the faceless nature of group chat allows people to speak without much thought to the consequence of their speech. Intonation and physical expression is lost and with it the subtle nature of human interaction."
6. The theme *Ineffective Tool; Doesn't Add Value* was in 5.3% of responses: "too complicated for the benefits it has" and "during the session, someone asked, 'Is anyone in the computer lab? If so raise your hand.' I think 80% of us were there within feet of each other. Just seemed pointless. You would think that is why you would use that technology, so you wouldn't have to leave your house, or go through all that trouble."
7. Ranked seventh, in 4.8% of responses, is *Disorganized Content.* Examples include: "it was hard to discuss things with so many opinions flying around and so many people getting off the subject" and "discussion threads and stuff like that can be difficult to follow."

What Could the Teacher Do to Make the Least Beneficial Technology Easier to Use?

When students were asked what teachers could do to help make the technology deemed least beneficial easier to use, five themes emerged.

1. The dominant theme, in 25.4% of responses, is *Student Training.* Many students admitted needing more training in how to use a technology: "my instructor could explain it to me," and "demonstrations, demonstrations, and more demonstrations."

2. Ranked second, in 14.5% of responses, is the theme *Encourage Use; Use in Class.* Some students want teachers to encourage or even require students to use the technology: "maybe just using it would be enough" and "make it mandatory, otherwise it is pointless."
3. The third-ranked theme, in 8.8% of the responses, is *Nothing.* These responses indicate that teachers need do nothing to make the technology easier; it is sufficiently easy to use: "nothing, it is already easy to use."
4. One way teachers could make the least beneficial technology easier to use, mentioned in 7.9% of responses, is to *Avoid the Technology*. This theme reflects the perception that nothing can make the technology more easy or effective, and thus it should not be used: "well, not using it would be a good start" and "abstain from using it, period."
5. Finally, the theme *Modify Technology* was mentioned by 5.3% of the students. The technology needs further development or modification, or students may offer solutions for perceived limitations: "offer an instant messaging system," and "maybe have COM ports on the desk so that we can sync up to the teacher."

Consistently, this research has failed to support the presence of a digital divide. We discovered no real differences across gender or racial/ethnic groups. However, we note that this may in part be an artifact of our sample; perhaps students whose teachers invite them to complete an online survey, and who are comfortable in doing so, have already been socialized into accepting technologies in a certain fashion, and have had enough consistent access that differences have disappeared.

Still, our results speak to "students" rather than any isolated group thereof, and underscore the primacy of e-mail. It seems as if our respondents couldn't survive in school without e-mail—or at least, they wouldn't want to try. No other technology even comes close to the ubiquity, comfort, and ease of use of e-mail. No other technology could be deemed vital in academe; e-mail certainly is.

PRACTICAL IMPLICATIONS FOR EDUCATORS

We conclude this chapter by presenting advice for teachers wishing to incorporate new technologies into the classroom.

- Of primary importance is the need for educators to provide sufficient training in all technologies used, and not to assume students know how to use technology, or that they know all aspects of the technology (e.g., many respondents indicated problems with e-mail attachments). This training should include not only technical skills but also the norms of technology use by both teachers and students (e.g., expected usage frequency, appropriate "tone of voice," range of acceptable content, general turnaround time for teachers to respond to students' e-mail, and the like).

- Educators must *be responsive*. Because students may mistrust the system when submitting assignments electronically, we recommend teachers provide an acknowledgment of receipt (even if automated). We also urge teachers to check e-mail regularly and respond in a timely fashion to student messages; given the importance of e-mail to students we should probably consider this a common courtesy.
- If teachers want students to use technologies, *teachers must use technology*, and encourage, perhaps even require, students to use it. Students learn by example, and teachers should model effective technology use. Further, teachers could increase use (and encourage appropriate use) of a technology by specifically showing the students how it will benefit them, and explicitly demonstrating how students' use of a technology will help them achieve course objectives.
- To use a technology well (and creatively), and to give students a good reason for using it, *most teachers will need additional training* themselves. This is a good arena for educators' training and development workshops. Teachers should learn from each others' valuable experiences.
- Don't use any technology just because it's there—*have a specific goal in mind* and monitor the extent to which it is accomplished. It need not be a lofty goal, but without a reason to do so, the technology should not be used.
- *Understand the physical limitations of access*; students often have to fight for space in public labs, or use slow dial-up modems from home. Teachers should not let students use access issues as excuses for poor performance, but rather should have a realistic understanding of the physical situation and prepare materials and expectations accordingly—facilitating successful and rewarding educational opportunities. If appropriate, educators may wish to urge their school administrations to provide additional computing support for students.

Finally, we urge educators to be sensitive when introducing technology, but not necessarily differently sensitive on account of gender or race/ethnicity. Provide adequate training, but don't assume that different groups have had (or will need) more training than will others. Remember that all students, regardless of gender, race, or economic background, have the right to achieve course objectives in a technological environment that is as relevant, flexible, egalitarian, mutually supportive and respectful, and user friendly as the instructor can provide.

NOTES

1. The authors wish to thank Joyce M. Fitzgerald (University of Illinois at Chicago) for her assistance in coding and data entry. The authors are also grateful to the educators who recruited their students to participate in the survey, as well as to the students themselves for so generously sharing their opinions.
2. Fuller descriptions of the themes, including additional examples, are available at *www.uic.edu/~rebecca/newtechreport.html*

REFERENCES

Barbieri, M.S., & Light, P. (1992). Interaction, gender and performance on a computer-based problem solving task. *Learning and Instruction, 2,* 199-213.

Besser, H. (1993). Education as marketplace. In R. Muffoleto & N.N. Knupfer (Eds.), *Computers in Education: Social, Political, and Historical Perspectives* (pp. 37-70). Cresskill, NJ: Hampton Press.

Brunner, C., & Bennett, D. (1998). Technology perceptions by gender. *Education Digest, 63,* (6), 56-58.

Busch, T. (1995). Gender differences in self-efficacy and attitudes toward computers. *Educational Computing Research, 12,* (2), 147-158.

Coffin, R.J., & MacIntyre, P.D. (1999). Motivation influences on computer-related affective states. *Computers in Human Behavior, 15,* (5), 549-569.

Cookman, C. (1998). A computer-based graphics course and students' cognitive skills. *Journalism and Mass Communication Educator, 53,* (3), 37.

Culley, L. (1993). Gender equity and computing in secondary schools: Issues and strategies for teachers. In J. Beynon & H. Mackay (Eds.), *Computers into classrooms: More questions than answers* (pp. 110-124). London: Falmer Press.

Gougeon, T.D. (1998, November). Gender sensitive instruction: A distance education issue. Paper presented at the American Association for Adult and Continuing Education, Phoenix, AZ. ERIC Microfiche ED 428 269.

Groves, M.M., & Zemel, P.C. (2000). Instructional technology adoption in higher education: An action research case study. *International Journal of Instructional Media, 27,* (1), 57.

Harrell, W., Jr. (1998). Gender and equity issues affecting educational computer use. *Equity & Excellence in Education, 31,* (3), 46-53.

Hawkins, R., & Paris, A.E. (1997). Computer literacy and computer use among college students: Differences in black and white. *Journal of Negro Education, 66* (2), 147-159.

Hess, F.M., & Leal, D.L. (1999). Computer-assisted learning in urban classrooms: The impact of politics, race and class. *Urban Education, 34,* (3), 370-388.

Huppert, J., Yaakobi, J., & Lazarowitz, R. (1998). Learning microbiology with computer simulations: Students' academic achievement by method and gender. *Research in Science and Technology Education, 16,* (2), 231-245.

Kantrowitz, B. (1994, May 16). Men, women, and computers. *Newsweek,* 48-55.

Kayany, J.M., & Yelsma, P. (2000). Displacement effects of online media in the socio-technical contexts of households. *Journal of Broadcasting & Electronic Media, 44,* (2,) 215.

Lewis, J., Whitaker, J., & Julian, J. (1995). Distance education for the 21st century: The future of national and international telecomputing networks in distance education. In Z.L. Berge & M.M.P. Collins (Eds.), *Computer-Mediated Communication and the Online Classroom,* Vol. 3. (pp. 13-30). Cresskill, NJ: Hampton Press.

Malveaux, J. (1997). Technology necessary but not sufficient for parity. *Black Issues in Higher Education, 14* (7), 34.

Margolis, J., Fisher, A., & Miller, F. (1999/2000). Caring about connections: Gender and computing. *IEEE Technology and Society Magazine, 18,* (4), 13-21.

McDowell, E.E., & Schuelke, L.D. (1998). An investigation of the relationships among technology experiences, communication apprehension, writing apprehension and computer anxiety. ERIC Microfiche ED 425 485.

Miller, M.D. (1996). Gender and undergraduate business students' perceptions of computer self-efficacy. *Psychological Reports, 79,* (3), 946.

Novak, T.P., & Hoffman, D.L. (1998). Bridging the racial divide on the Internet: Access along racial, income and educational lines. *Science,* 280, 5362, 390-391.

Paneitz, B. (1997). Community college students' perceptions of student services provided when enrolled in telecourses. (Doctoral Dissertation—University Microfilms No. 9735008).

Pingree, S., Hawkins, R.P., Gustafson, D.H., Boberg, E., Bricker, E., Wise, M., Berhe, H., & Hsu, E. (1996). Will the disadvantaged ride the information highway?: Hopeful answers from a computer-based health crisis system. *Journal of Broadcasting & Electronic Media, 40,* 331-353.

Raloff, J. (1998). Internet access: A black and white issue. *Science News, 153* (16), 247.

Robertson, I., Calder, J., Fung, P., & O'Shea, T. (1995). Computer attitudes in an English secondary school. *Computers and Education, 24,* 73-82.

Schumacher, P., & Morahan-Martin, J. (2001). Gender, Internet and computer attitudes and experiences. *Computers in Human Behavior, 17,* (1), 95-111.

Spotts, T.H., Bowman, M.A., & Mertz, C. (1997). Gender and use of instructional technologies: A study of university faculty. *Higher Education, 34,* (4), 421-436.

Underwood, J.D.M., & Geoffrey, U. (1990). *Computers and learning: Helping children acquire thinking skills.* Cambridge, UK: Basil Blackwell.

U.S. Department of Commerce National Telecommunications and Information Administration. (2000). Falling through the Net: Toward digital inclusion. Online at http://www.ntia.doc.gov/ntiahome/fttn00/contents00.html

WEEA Digest: Beyond equal access. (2001). Online at http://www.edc.org/WomensEquity/pubs/digests/digest-beyond.html

Weinman, J. & Haag, P. (1999, February). Gender equity in cyberspace: Gender differences in the use of computer-based technologies. *Educational Leadership, 56,* (5), 44-50.

Zehr, M.A. (2001, May 10). Rural Connections. *Education Week.* Online at http://www.edweek.org/sreports/tc01/tc01article.cfm?slug=35rural.h20

CHAPTER 11

Analog-to-Digital Conversion Considerations

Bruce Mims

INTRODUCTION

When compact discs (CDs) first began to appear in radio station control rooms around 1984 (McGinley, 1984, p. 14), how many broadcasters envisioned the beginning of a revolution in the way that audio would be created, manipulated, stored, and transmitted? Perhaps quite a few had no difficulty foreseeing the imminent displacement of the turntable and vinyl record by CDs. After all, the phonograph for playing flat-disc records, which Emile Berliner introduced in 1893, had evolved and matured into its high-tech, direct-drive, quartz speed-controlled state by the early 1980s (Baert, Theunissen, Vergult, Maes, & Arts, 1995, pp. 1-3). It nonetheless relied on an electromechanical design perfected in the mid-1920s (Alkin, 1981, p. 144).

Magnetic tape recording technology, on the other hand, was a different story. In the 1980s, broadcasters viewed tape as a technology of indeterminate shelf life. More electronic in form and function than the phonograph, this method of audio storage and retrieval earned its reputation as the broadcaster's servant. The Otari Corporation deservedly advertised their decks as the "workhorses" of two-channel analog tape recording (Otari, 1989, p. 13). Open-reel and cartridge tape machines manufactured by Otari as well as others performed yeoman service for broadcasters, assisting air staffs in managing a host of duties ranging from news acquisition to spot playback to automated program origination.

In spite of late-1980s forecasts, the future of digital audio technology as the replacement for analog tape-based audio production was uncertain. In 1988, one industry observer predicted the future of a "cost-effective, user-friendly, all-in-one digital audio production tool" to be five years ahead (Pizzi, 1988, p. 30). The following year, Michael Rau, Vice President of

Science and Technology for the National Association of Broadcasters (NAB), proclaimed the future of digital audio production to be "murky," its techniques hampered by a "hodgepodge" of technical standards (1989, p. 39). For these reasons it was difficult for any seasoned veteran to foresee a day in which nowhere in the station could a cart machine, splicing block, or empty take-up reel be found. Yet, within a decade of the CD's introduction, evidence of a paradigm shift in audio technology had begun to accumulate. Digital audio sounded better than analog and it was more cost-efficient ("Equipment: Riding," 1991, p. 44).

The arrival of digital audiotape (DAT) decks in the mid-1980s signaled the beginning of the shift in audio recording. Originally intended for consumer use, DAT found its way into the broadcast control room due to its superior sonic performance and its ability to make copies of recordings that were indistinguishable from the original productions (Braverman, 1989, p. 187). Producers and engineers soon began to regard the DAT format, with its editing difficulties and frequent deck maintenance requirements (Alten, 1999, pp. 121–122), to be the adjunct to analog recording rather than its replacement. Although the broadcast industry's response to DAT was less enthusiastic than manufacturers had hoped, this medium afforded many recordists their first opportunities at audio production in the digital arena.

The 1990s was a pivotal decade in terms of technological advancement in audio production. Both broadcasters and broadcast educators entered the decade with a solid fix on radio's analog past but also an optimistic vision of its digital future. With the power of hindsight guiding their interpretation of advancements during that period, present-day industry observers might assert that analog audio had run its course by the end of the 1990s. A new generation of devices for multitrack recording, editing, and playback introduced to radio stations during that decade affirmed the supremacy of audio digitization (Rowland, 1994, p. 17). Software developers harnessed the power of the personal computer (PC) in creating a new breed of programs for manipulating audio signals, sporting features such as nondestructive editing, multiple levels of undo, and immediate location access (Jones, 1994, p. 54).

Automated broadcasting equipment similarly made the transition from analog tape to computer hard disk storage. Broadcasters and listeners alike criticized the first-generation automation systems of the 1950s and 1960s, which relied entirely upon analog recording/playback devices, for creating programming that was bland and lifeless (Antilla, 1981). PC-based automation, alternately, quickly earned a reputation for enabling station personnel to create the illusion of the live, spontaneous broadcast with easily produced and realistically sounding prerecorded voice tracks. In contrast with "canned"-sounding, tape-based automation systems, PC automation dared listeners to differentiate between live and voice-tracked presentations (Zarecki, 2001).

Speculation about reasons why small-market broadcasters became the early adopters of digital editing and automation technologies is in far greater abundance than are actual, documented reasons. Reist (1993) asserted that managers of small-market stations stood to benefit from making the conversion to digital because of the labor supply. College graduates of that era typically exhibited greater proficiencies with computers than did veteran employees of stations in the entry-level markets (p. 12).

Other plausible explanations include one offered by Jay Mitchell, author of the *Small Market Radio Newsletter.* He observed that "small market stations in general tend to be early adopters of technology in general" In contrast with major market stations, Mitchell noted that stations in small markets "are able to get things up and running a lot faster . . . since the larger ones are more bureaucratic and think in terms of outsourcing more than their smaller brethren" (Mitchell, 2000). On a related economic note, hard disk recording and editing by the mid-1990s became price-competitive with comparable-quality analog tape decks (Gross, 1994, p. 21). Also creating concern for station managers was a general decline in engineering talent available to service and maintain hardware (Martin, 2000, p. 50).

Broadcast educators must wonder if valid reasons exist for continuing to offer courses to students whose purposes are to assist them in developing their proficiencies in the use of obsolescent, analog technology. For instance, instructing students today in the techniques of splicing magnetic recording tape or dubbing audio onto tape cartridges may be regarded as exercises in technological futility. If educators make uninformed decisions about equipment or adopt nonstandard technologies, they risk creating facilities ill-equipped for their purposes. Depending upon the availability of funding, it could take years to correct imprudent decisions. In instances where strong correlations between curriculum and specific items of equipment exist, such as those found in production-techniques classes, poor decisions could also result in unplanned revisions to course content. Clearly, there is a compelling reason for ensuring that decisions involving technology are consistent with those being made in the broadcast community.

The purpose of this chapter, therefore, is to assist broadcast educators, administrators, and station advisers in the difficult processes of identifying the issues surrounding technology and its integration into the curriculum. In reporting on the opinions expressed by broadcast professionals about technology and equipment, the chapter offers insight for academicians into the dual processes of selecting technologies and purchasing equipment that conform to the selected technological standard(s).

HIGHLIGHTS OF PREVIOUS STUDIES

For a period of more than 60 years, from the inception of radio broadcasting until the early 1980s, studio production equipment was exclusively of analog design. Save for the impact that magnetic tape recording had on the industry's production techniques in the late 1940s, radio's first and only genuine technological transformation to occur is its current migration to digital. Because digitization in the radio studio is largely a phenomenon of the 1990s, it must be emphasized that the body of literature that addresses this chapter's topic is relatively small in size and narrow in its focus.

In several instances, a common thread that weaves through recent studies is the question (or a related set of questions) inquiring about the methods for teaching production skills to students. Broadcast educators and administrators question the extent to which the content of production courses should reflect the changes within the industry. It is a situation that creates uncertainty not only about what should be taught but also how it should be taught. Among the important findings are the indications suggesting that educators carefully reevaluate their curricula and approaches to instructing students in the operation of audio production equipment during this period in which the industry redefines its technological platforms.

Keith (1998) underscored the need for collaboration between members of the profession and the academy in a *Journal of Radio Studies* forum involving four nationally recognized radio broadcasting executives. In asserting the need for educators to fine-tune curricula to resonate with current industry practices, Norman Pattiz, Founder and Chairman of the Board of Westwood One Communications and former president of the Broadcast Education Association, observed a need for improving the lines of communication and the exchange of information. Pattiz observed,

> In every college community, there are commercial radio stations. Why not create local professional/academic alliances? These partnerships could provide access to professional instructors, student internships, and maybe even opportunities for funding and industry support. The resources are there. The student interest and industry demand are there. The opportunity is there. (pp. 6–7)

In one of the earliest studies to emerge during the digital revolution, Brown (1991) examined the pedagogy/technology issue, noting that advisers of student media production facilities could be trapped in a curriculum/equipment conundrum: in addition to supervising the content and execution of programming, they were often expected to specify and oversee the acquisition of studio equipment. Advisers who serve in this dual capacity must ensure that in instances where technological standards are concerned, only the most informed decisions are made. A comprehensive understanding by advisers of the technologies used by broadcasters offers the best insurance that the production skills students acquire are the ones consistent with those expected by professionals.

Making purchase decisions, however, can create multiple problems for advisers, particularly because the process compels them to both stay abreast of developments in the technology and integrate this information into classroom presentations and laboratory assignments. Just as their counterparts in the profession, broadcast educators must be able to evaluate new technologies in terms of their likelihood for acceptance or rejection by the industry. They must determine when investments into new technologies should be made and, perhaps more importantly, to what extent they are to be utilized. Advisers must also be mindful that there will come a time in which the expense of maintaining and repairing existing equipment surpasses the cost of its replacement. The process of determining when equipment reaches the end of its useful life must be factored into the equation when requests for replacing obsolescent technologies are made to administrators.

TRANSITION CONSIDERATIONS

This discussion raises several issues for consideration. One addresses the concerns that surround the analog–digital debate: because the industry continues to support an intermixture of technologies, should curricula for production courses reflect this situation? If so, into what proportions should class and laboratory times be devoted to the discussion of each technology? Furthermore, in instances where stations within an institution's service region employ digital audio technology exclusively, educators must determine which software/platform combination will offer students the greatest level of preparation.

Prior to the arrival of digital audio, decision making was a much simpler task. For example, open reel analog tape decks, regardless of their origin, operated in essentially the same manner. Can the same be said for the operation of digital audio workstations? Mastering the techniques of audio editing in the digital domain can become a much more intensive process because some of the more popular workstations have signal-processing features layered on top of the recording, playback, and editing functions. For example, internal system control over the equalization, compression, and peak limiting of the audio waveform is commonly configured into editing software. Systems that feature built-in digital signal-processing power eliminate the need for relying upon outboard devices in the studio for controlling these parameters, but they also increase the amount of time needed to achieve mastery.

On a related note, broadcast educators also must be aware of employer expectations about the production skills recent graduates possess. For example, should entry-level employees be product-specific in their knowledge of audio editors? Conversely, are employers prepared to hire producers with nothing more than a rudimentary knowledge of editing concepts and who must be trained in the use of station-specific equipment? Interestingly, this issue is one that also deserves examination from the student perspective. Is it reasonable for a student to expect to receive instruction in the use of the brand(s) of software used most frequently by stations within the institution's service region?

By the conclusion of the 1990s, manufacturers of audio broadcast equipment had achieved essentially total digitization of the program audio chain. Nonetheless, even in markets as large as New York City, analog production equipment continues to coexist with machines incorporating more recent digital technology (Hausman, Benoit, & O'Donnell, 2000, pp. 14–15). The cartridge tape ("cart") format best illustrates this assertion. Shortly after its introduction in the late 1950s, the cart streamlined control room operations. With its endless loop of recording tape tucked neatly inside a plastic package, the cart represented a dramatic improvement over open-reel tape, thanks to its ease of use and its self-cueing ability. Announcers no longer had to deliver commercial copy live; the cart format enabled them to record announcements prior to their broadcast. When it came to production value and listener appeal, "live-read" advertising messages paled in comparison with recorded commercials because producers could embellish recorded vocal deliveries with music and sound effects. All sorts of audio messages, ranging from station jingles to public service announcements to news actualities, adapted easily and readily to playback from cart, enhancing the flow and pace of the broadcast (Mims, 1998). Adams and Massey (1995) acknowledged the ubiquitous nature of the cart in observing:

> Those five-, ten- and thirty-second production aids that give radio its *sound* make the cart still the most popular piece of on-air equipment. Add to that the advantage of being able to record any locally produced audio onto it, and you can see why it has been an industry standard since 1960. *Even digital replacements for the cart have been made to look and feel like the familiar cart* [emphasis added]. (p. 36)

There is a growing consensus among broadcasters and broadcast educators that the cart has reached the end of its life cycle. Several digital recording/playback devices, including the MiniDisc, the Short/cut, and the DigiCart, as well as software products such as CartWorks, all have demonstrated their ability to equal or surpass the performance of analog cartridge tape. Yet carts retain their foothold in some of the most prominent stations in the largest markets. A person need spend only a few minutes viewing MSNBC's telecast of the nationally syndicated program *Imus in the Morning* from New York City to understand flagship station

WFAN's continuing commitment to cartridge tape. And a few minutes spent observing the control room configuration of the NBC television program *Frasier* reminds us that art imitates life when it comes to the venerable cartridge. Industry observers predicted in mid-2001 that cartridge tape would likely be serving as the principal audio storage/reproduction technology at some stations for the next five years (Fybush, 2001, p. 10).

Studio environments in which analog and digital technologies commingle beg the question of how much longer the techniques of analog audio production should be taught. Research into this situation is sparse and the results are inconclusive. Three recently conducted studies help to illustrate this important point. A recent snapshot view of the efforts by state-supported institutions of higher education in Pennsylvania to maintain the technological currency of their student radio stations is offered by Wilson and Dudt (2001). Results of their survey of the Keystone State's 13 facilities operated at institutions governed by the State System of Higher Education (SSHE) indicate that eight stations continue to operate with analog facilities of varying degrees of equipment condition. Five stations report that the conversion to digital technology either has been completed or is under way.

Two pieces of information reported in the survey results bear noting. First, and most important, is the fact that desires by station managers to convert are not met with equal levels of institutional support. Funding mechanisms for supporting equipment purchases range from one-time special allocations that enable complete conversion to piecemeal disbursements for single-item acquisitions. Also to be noted is the fact that, with one exception, SSHE stations operate under authority of Student Government Associations and are not integrated with academic courses of instruction (pp. 8–9).

Blaney and Donnelly (2000) conducted focus-group interviews with department chairpersons of Broadcast Education Association-member institutions. These researchers elicited comments from the participants about the state of the industry's technological advancement and its impact on student instruction. The consensus opinion expressed by administrators, Blaney and Donnelly reported, was that "the acquisition of digital production skills far outweighs those of analog skills because in recent years the radio industry has become almost entirely digital" (p. 4).

This may be an instance in which perception and reality diverge. Lind and Danowski (1999a) question whether the industry and the academy are tuned to the same wavelength when it comes to preparing students for entry-level employment. Survey responses offered by 303 broadcasters and an approximately equal number of broadcast educators led the researchers to an indefinite conclusion about the content and instruction that educators include in their courses. The uncertainty resides with either how liberally or how conservatively readers interpret Lind and Danowski's finding on students' computer-competency expectations. They report that "broadcasters place significantly more weight on computer skills than do educators. Computer skills is as technical as broadcasters get, while educators focus on technical and production skills" (p. 2).

Whether knowledge of "production skills"—as defined by educators—is subsumed within the definition of the "basic skills" broadcasters expect of graduates remains an unanswered question. Lind and Danowski interpret the findings in such a way as to suggest that broadcasters attach less importance about graduates' knowledge of production skills than do educators (p. 3). One approach to resolving the knowledge gap between the academy and the industry, they assert, is to open the lines of dialogue: "Increased communication between broadcasters and educators on this issue," they concluded, "can resolve whether the difference is real and if so, educators can factor this into their curriculum decisions. *For some educational institutions it would be a large cost savings not to have to upgrade equipment* [emphasis added] (Lind & Danowski, 1999c, p. 1).

MISSOURI BROADCASTER STUDY

Coinciding at the approximate time of the data-gathering stage of the Lind and Danowski study, the author of this chapter was also gathering radio broadcaster opinions about graduates' production skills and knowledge. In contrast with the nationwide scope of Lind and Danowski's effort, this author directed questionnaires to members of a statewide association of broadcasters in the Midwest. Readers should be mindful of two facts associated with this study. First, it should be noted that all survey respondents resided in the same state. Therefore, the extent to which its results may be generalized to the national population of broadcasters is undetermined and inconclusive. The second consideration is the age of the data. Manufacturers of computers and digital audio production equipment continue to refine and improve the processing speed and storage capacity of their products at an amazing pace. Devices that exhibited cutting-edge performance just a couple of years ago may be viewed today as commonplace.

Analog audio recording, on the other hand, has been regarded as a mature technology for a number of years. Its longevity can be attributed to its robust reliability and to an industry of producers and engineers who were weaned on products bearing the names Ampex, Scully, Gates, ITC, and Otari and of other reputable manufacturers of reel and cartridge tape decks. In the ensuing discussion, the author interprets the data from a perspective that acknowledges conditions both as they existed at the time of data collection in the late 1990s and, with regard for the advancements that have occurred since then.

While the survey reported in this chapter was conducted independently of the efforts of Lind and Danowski, it nonetheless reinforces their recommendation for educators to actively engage professionals in discussions about curricula and technology. Its success in achieving this purpose should be evaluated in terms of the responses submitted by its participants. The pattern of affirmative responses to emerge from the results indicates that broadcasters are actively interested in the impact of technology on broadcast education issues. When station personnel make employment decisions, they expect recent graduates to possess knowledge of specific equipment types and be able to exhibit specific competencies when it comes to the use of that equipment.

METHOD

Select members of the Missouri Broadcasters Association (MBA) were invited to participate in the survey. Questionnaires were mailed to all member-managers of MBA-affiliated radio stations utilizing mailing address labels provided by the executive vice president of the association. Initial mailings of the survey instrument, titled "Equipment College Graduates Should Know How to Operate," occurred in September 1998. Managers (or their designees) were asked to return the questionnaires within the next 30 days. Follow-up mailings were posted in October of that year and the data were tabulated the following November.

Data were obtained by means of a five-step Likert scale on which members could record responses to questions about equipment technologies, brands, and production processes. Scale levels asked participants to respond by selecting one of five options: (1) "very important," (2) "important," (3) "somewhat important," (4) "not very important," or (5) "not at all important." Respondents also were given the option of supplying the brand names/models of equipment not identified specifically within the questionnaire.

The MBA office provided 112 useable address labels. Nine labels were rejected because of duplication of recipient names and stations, presumably the result of station ownership-consolidation changes that had occurred too recently to be reflected in the MBA database. MBA members returned 43 useable responses. Approximately three-quarters of the respondents (34 out of 43) identified themselves as station managers. Directors of programming, operations, and engineering departments combined returned nine instruments.

A return rate of 38.4% was achieved, which falls within the 30 to 60% rate characteristic of mailed questionnaires (Comstock & McCombs, 1981, p. 146). Yu and Cooper (1983) report that a response rate of 47.3% is typical for mail surveys conducted by social scientists (p. 39). By way of comparison, the industry publication *BE Radio* achieved a surface-mail response rate of 36 percent in its 2000 Salary Survey of the magazine's subscribers (Martin, 2000, p. 44).

It should be noted that the response rate falls below the 50% benchmark that social scientists regard as desirable for ascribing reliability to studies (Babbie, 1992, p. 267; Tucker, Weaver II, & Berryman-Fink, 1981, p. 96). An important presumption about the survey response rate has been made. It relates to the size of markets within the state of Missouri, which range from single-station communities to the municipalities of St. Louis and Kansas City. St. Louis, with 30 stations serving the metro listening area, and Kansas City, with 27 stations (Broadcasting & Cable, 1999), are classified as "large" markets (O'Donnell, Hausman, & Benoit, 1989, p. 3).

Although respondents conveyed their opinions anonymously, they were asked to indicate the number of signals that originated from within their market. Their responses to this question ranged from 1 signal to 27 signals, offering relative indications of the market sizes from which questionnaires were returned. Only one questionnaire indicated origination from a market of 21 or more signals, which leads this author to the assumption that the vast majority (42 out of 43) of respondents were located at small- and medium-market stations. Small and medium markets offer more entry-level employment opportunities than do large markets (Lind and Danowski, 1999b, p. 1). Smith, Wright II, and Ostroff (1998) offer a more blunt assessment of the relationship between market sizes and career opportunities for beginners: "First, you start at the *career entry level*—the place where jobs are (a) available (b) for beginners. That place is not the networks, . . . not a big market, and not a big station" (p. 566). In view of the likelihood that few, if any, recently graduated persons would be able to secure employment in a program-production capacity in either St. Louis or Kansas City, this author presumes that MBA members in St. Louis and Kansas City elected out of participation in the survey, thus lowering the anticipated response rate.

QUESTIONNAIRE TOPICS

The three-part questionnaire began with an introductory section in which respondents were asked to identify their position within the station (or station cluster) and to record the number of stations serving that market. It began by focusing on specific items of analog and digital equipment commonly utilized in the on-air and production studio environs. This part of the instrument asked respondents to rate the importance of student knowledge about dedicated analog and digital record/play decks. Part II of the questionnaire invited them to express opinions regarding students' proficiencies in the use of digital editors and digital program-automation systems. Companion questions in Part III offered opportunities for rating editing and automation by brand name and/or model.

INTERPRETATION OF SURVEY DATA

Digital Audio Technologies

This portion of the survey instrument asked station managers to respond to three related questions about digital audio technology. Each question was constructed to focus on the knowledge they would expect of recent graduates with respect to two items of equipment that are becoming increasingly more important to the operation of radio stations: digital editors and PC-based program automation systems. The category of digital editors was subdivided into two categories to account both for systems that operate on computer platforms as well as those that function as stand-alone units, operating under control of their own internal systems. Results of the responses to these questions are indicated in Table 11.1.

The most revealing data to emerge from the results are those pertaining to program automation. Thanks to advancements in PC reliability and to significant increases in hard-disk storage capacity, program automation systems are now in widespread use. Matching the power of today's PC-based, voice-track friendly automation systems with the talents of a local announcer or of one imported from outside the market results in a product that is virtually indistinguishable from live operation. When the techniques of automated broadcasting are executed properly, listeners have almost no reason to question whether the presentation is live or recorded (Zarecki, 2001).

In acknowledging the sophistication with which systems can intermix sound files of various program elements, it should not surprise educators to learn that almost 90% of the survey's respondents believe it is "very important" for graduates to be knowledgeable about digital studio

Table 11.1
Broadcaster Percentage Responses: Digital Audio Equipment

	Very Important (1)	(2)	(3)	(4)	Not Important (5)
1. PC-Based Digital Audio Editor	77	14	7	—	2
2. Dedicated Editing Workstation	51	28	19	2	—
3. PC-Based Control Room Automation	88	7	2	2	—

automation systems. Another explanation for such a high percentage response may be ownership consolidation, which has resulted in situations in several Missouri markets where licensees originate five or six signals from a central studio location. One respondent, whose stations compete in a 22-signal market, commented that graduating students' knowledge of studio automation was "incredibly important." Clearly, broadcasters are telling educators that knowledge of automation-system operation is one of the most highly desired skills they expect new hires to possess. Their system of preference is Broadcast Electronics' *AudioVault*, which was ranked "most important" by 29% of the respondents. Scott Studio's *NT System* and Radio Computing Service's *Master Control NT* followed with 17% and 11%, respectively.

Integrating automation technology into the curriculum raises a couple of issues, however. One of these is expense. As is the case for most PC-related products, prices for hard-disk automation systems have decreased steadily in recent years, making their purchase feasible for all but the most budget-meager institutions. Entry-level pricing in 2001 for the top-ranked BE *AudioVault*, for example, is less than one-half of the system's cost just four years earlier, according to a company representative (J. Stratton, personal conversation, May 2, 2001). Similar percentage decreases in the cost of Enco's DADPRO 32 and Radio Computing Systems Master Control systems have occurred.

A concern of greater importance than expense is that of pedagogy: what concepts and skills associated with automated radio broadcasting should be incorporated into the curriculum? Since 1958, when Paul Schafer introduced his first audio program controller (Gentry, 1988), the radio industry has ridden several waves of infatuation with program automation systems. The electromechanical, analog systems of the 1960s fell into disfavor because their execution of the sequence of program elements conformed to predictable patterns. Microprocessor-controlled, tape-based systems of the 1980s revived interest for a period of several years, but by the end of the decade, interest waned for similar reasons.

The real revolution began to unfold in the 1990s when manufacturers married microprocessor control techniques with digital audio storage. For the first time, true random accessibility over recorded program elements had been achieved. The ease with which announcers could record, edit, and intersperse their comments within other program elements led them to create shows that sounded not only *live,* but also *alive.* Automation enables announcers to record, or *voice track,* the entire spoken portion of the typical four-hour airshift within an approximately fifteen-minute period, freeing them from the confines of minute-by-minute studio operations to pursue other creative endeavors for their employers. Operators of automated stations could lock the studio doors at the completion of weekday office hours, trusting their *AudioVault, Master Control,* or similar system to execute the prerecorded overnight and weekend airshifts with minimal operator assistance or intervention.

Voice tracking is an art in and of itself. Announcers who are otherwise articulate, spontaneous, and witty during live air shifts might find it difficult to summon those talents on demand and convey them within the condensed time frame in which the comments are created and stored into the system. Because the demands placed on tomorrow's professionals most likely will include responsibilities in automated control rooms, the importance of emphasizing the development of voice tracking skills in the classroom and laboratory station cannot be over-emphasized. The obligation for educators is clear. Programs of instruction must recognize voice tracking as a unique announcing skill independent of that required of announcers during live operation. The method presents its own set of challenges and requires students to exhibit mastery in accordance with the unique demands it places on the operation of stations. Its impact and influence on control room procedures suggest that advisers of student-operated laboratory stations consider converting schedules from live operation to either live-assist or fully automated modes in order to expose students to the studio environment they surely will encounter as professionals.

Digital Audio Editing

Broadcasters similarly expect graduating students to have reasonable familiarity with digital audio editors. More than three out of four (77%) of the respondents said that it was very important for prospective employees to be knowledgeable in the use of PC-based digital audio editors. This percentage contrasts starkly with the responses reported in Table 11.1 about the analog equivalent for audio editing, the open-reel tape deck. The fact that respondents favored digital over analog by an almost two-to-one margin might at first suggest that educators seize the future and abandon instruction in the use and care of analog tape decks. The high percentage of responses assigned to the "very important" criterion reinforces the recommendation offered by respected industry professional Norman Pattiz at the beginning of the chapter.

The survey instrument invited respondents to identify by brand the software editing package they would expect graduates to be able to use. In addition to identifying three of the more widely advertised packages, the instrument also provided an opportunity for respondents to add unlisted brands. Administrators and advisers involved in making equipment purchase specifications should note that survey respondents made Innovative Quality Software's SawPlus editor a strong favorite among the identified software brands (42% considered it "very important") for PC-based editing. Other software brands listed by respondents included Cool

Edit Pro, Digital Concepts, and Fast Eddie, although each received less than 10% of the respondents' rankings as "very important."

Missouri broadcasters were slightly less enthusiastic about dedicated digital audio workstations. One explanation for this, insofar as products such as Orban's Audicy is concerned, could be attributed to their cost, which was several thousands of dollars higher than comparable PC-based systems at the time of the survey. Similarly low rated was the more affordable dedicated-function editor, the Roland 1680. Although one out of two (51%) station personnel considered it "very important" for students to able to operate a dedicated digital editor, no product emerged from the responses as a clear-cut favorite. A lack of consensus among respondents in this category is underscored by the knowledge that only one editor—the Audicy—elicited more than a single-digit favorable response percentage. Also, the fact that dedicated editors function under control of proprietary operating systems, as opposed to software-based editors that operate on the more widely utilized PC platform, may account for some of the reluctance of broadcasters to invest in these products. While manufacturers continue to offer dedicated editors for sale, the number of models has decreased since the time of the survey's completion. Insofar as the survey reveals that radio broadcasters respond more favorably to software-based editing programs, educators should carefully examine the advantages and disadvantages of proprietary systems before making financial commitments. A few hours of time spent investigating the state of this technology at the time of the purchase decision could go a long way toward minimizing the risk of standardizing on an operating system that has not and will not achieve widespread acceptance by the industry. An additional consideration, Maycock (2000) notes, is that computer-based editing stations typically afford much greater flexibility in changing operating parameters. The ability to reconfigure an editor's input/output options is more easily accomplished with PC-based digital audio workstations (p. 66).

Record/Play Decks

Given the strong level of interest that respondents expressed for digital technologies, one might expect analog equipment to have fared poorly in the survey. To the contrary, broadcasters indicated that knowledge of the operation of analog record/play decks is of roughly equal value to knowledge of their digital counterparts. Table 11.2 data reflect the responses of broadcasters to seven types of analog and digital record/play decks.

Notable among the "very important" responses were broadcasters' indications that knowledge of the operation of analog open reel and cassette decks remained a desirable skill for entry-level personnel to possess. Knowledge of PC-based CD recorders was the only category to receive a higher percentage response. Digital audiotape (DAT) and Minidisc (MD) technologies received smaller percentage ratings than did either open reel or cassette tape decks. These data suggest that Hausman et al. (2000) are correct in asserting that students will continue to encounter analog recording technologies in the industry and that knowledge of the operation of analog decks continues to be an employer expectation (p. 15).

Table 11.2
Broadcaster Percentage Responses: Record/Play Decks

	Very Important (1)	(2)	(3)	(4)	Not Important (5)
1. Cartridge Tape Deck	26	2	21	26	26
2. Reel-to-Reel Tape Deck	42	5	26	16	12
3. Cassette Tape Deck	42	12	14	26	7
4. Digital Audiotape (DAT) Deck	40	28	26	2	5
5. Minidisc (MD) Deck	28	28	19	16	9
6. Stand-Alone CD Recorder	44	19	14	12	12
7. PC-Based CD Recorder	51	23	12	7	7

CONCLUSION

What use do educators make of this knowledge? For the near future, it appears the radio station production facilities will continue to host an intermixture of analog and digital technologies. Certain facilities, located in markets large and small, cling to analog audio recording because the technology has a demonstrated low failure rate when the equipment is properly serviced. Additionally, stations continue to support analog because the transition to digital involves significant capital expense.

For broadcast educators, the problem this hybridization of studio technologies presents is in determining how to apportion funds for the purchase and maintenance across the two technologies. It might not be difficult for institutions to justify the expense of *maintaining* existing analog decks, but do the opinions of surveyed broadcasters also suggest the need to invest in the purchase of *new* analog equipment? Advisers to student radio stations and production facilities acknowledge the difficulties faculty members face in efforts to balance academic obligations and expectations while also attempting to process information about the broadcast industry's technological upheaval.

A couple of examples remind us of how quickly change is occurring. Consider the technology of radio news gathering. When this chapter's inquiry about technologies was conducted in 1998, the analog audiocassette already had been surpassed by Sony's MiniDisc format. When Sony introduced the MD format in the mid-1990s, the electronics manufacturer targeted MD, like its audiocassette predecessor, toward consumer rather than professional applications (Maunez-Cuadra, 1995, p. 19). Yet a majority of Missouri broadcasters, when posed the question, hold favorable views about MD technology. Survey responses, in fact, revealed that MD already had gained acceptance as the preferred recording format for field acquisition.

At the time of this writing, it appears that the industry has taken the next step with digital recording. Electronics manufacturer Marantz, whose portable cassette decks set the standards for professional cassette recording, currently offers a deck that operates on the flash memory principle. Unlike both the audiocassette and the MD decks, the Marantz PC Card Recorder has no moving parts (Lindemann, 2001). The question that remains unanswered is whether this feature, as unique and desirable as it may be, warrants its considerable monetary investment. Does this technology represent an advancement of such magnitude that it warrants rendering otherwise acceptably performing technologies into obsolescence?

Evidence of rapid change is also observed in the area of program automation hardware and software. While it is possible today to purchase program control software from a variety of vendors and install it in off-the-shelf computers, such has not always been the case. At the time the Missouri survey was conducted, manufacturers of the three leading automation systems (Broadcast Electronics, Radio Computing Systems, and Scott Studios) required customers to purchase manufacturer-specified and -supplied computers and sound cards. From an administrator's perspective, it is well worth noting that the entry-level cost of automating a radio station can be achieved today for considerably less expense than just a couple of years ago—and with only minor sacrifices in hardware performance reliability.

This chapter offers information that station advisers and administrators alike can apply when making decisions about broadcast audio technologies and the instructional methods that underpin them. Additionally, it offers information to guide administrators during budget-request and equipment-procurement processes. The comments and opinions about equipment expressed here by media professionals can assist advisers in administering their allocations wisely, especially by helping to prevent investment of time and money into a technology or brand/model that has not achieved widespread adoption by the industry.

Efforts such as the survey of Missouri broadcasters reported here opens the lines of communication between the academy and the industry and yields a better understanding of production-technique advancements. Perhaps more importantly, however, the dialogue reinforces the need to educate students about media convergence—the problems it presents and the promise it affords. The acquisition and distribution processes will result in audio products of greater and more dynamic proportions. As the lines that distinguish media forms and formats continue to blur, the obligation for educators and practitioners to assess the audio production process increases. Collectively, our responsibility is not to determine "if"—but "how"—it can be achieved.

REFERENCES

Adams, M. H., & Massey, K. K. (1995). *Introduction to radio: Production and performance.* Dubuque, IA: Wm. C. Brown.

Alkin, G. (1981). *Sound recording and reproduction.* London: Focal Press.

Alten, S. R. (1999). *Audio in media* (5th ed.). Belmont, CA: Wadsworth.

Antilla, S. (1981, March). Canned radio means fresh profits. *Dun's Review,* pp. 104–106, 108.

Babbie, E. (1992). *The practice of social research* (6th ed.). Belmont, CA: Wadsworth.

Baert, L; Theunissen, L., Vergult, G., Maes, J., & Arts, J. (Eds.). (1995). *Digital audio and compact disc technology* (3rd ed.). Oxford, England: Focal Press.

Blaney, J., & Donnelly, G. (2000). Relationships between academic broadcast facilities and pedagogical outcomes. *Feedback, 41*(1), 1–8.

Braverman, D. (1989). RS adapts DAT for on-air use. *Radio World annual.* Falls Church, VA: Author.

Broadcasting & Cable (1999). *Broadcasting and cable yearbook 1999.* New Providence, NJ: Author.

Brown, Jr., L. W. (1991). Equipment broadcast students need to know. *Feedback, 32*(3), 21, 26.

Comstock, G., & McCombs, M. E. (1981). Survey research. In G. H. Stempel III & B. H. Westley (Eds.), *Research methods in mass communication* (pp. 144–166). Englewood Cliffs, NJ: Prentice-Hall.

Equipment: Riding out of a recession valley on a digital horse. (1991, September 9). *Broadcasting,* pp. 42, 44–46.

Fybush, S. (2001, April 25). Cart guys keep the format alive. *Radio World,* pp. 10, 12.

Gentry, R. (1988, April). Profile: Paul Schafer. *Broadcast Management/Engineering,* pp. 60-65.

Gross, J. (1994, January 12). A new year's wish list for stations. *Radio World,* pp. 21, 24.

Hausman, C., Benoit, P., & O'Donnell, L. B. (2000). *Modern radio production: Production, programming, and performance* (5th ed.). Belmont, CA: Wadsworth.

Jones, B. (1994, April 20). There are no DAW demons to fear. *Radio World,* p. 54.

Keith, M. C. (1998). JRS forum. *Journal of Radio Studies, 5*(1), 1–8.

Lind, R. A., & Danowski, J. A. (1999a). Executive summary. *Training tomorrow's broadcasters: A comparison of employers' needs and educators' intentions* http://www.uic.edu/~rebecca/educrec.html

Lind, R. A., & Danowski, J. A. (1999b). Methods. *Training tomorrow's broadcasters: A comparison of employers' needs and educators' intentions* [Online]. Available: http://www.uic.edu/~rebecca/educrec.html

Lind, R. A., & Danowski, J. A. (1999c). Recommendations. *Training tomorrow's broadcasters: A comparison of employers' needs and educators' intentions* http://www.uic.edu/~rebecca/educrec.html

Lindemann, C. (2001, March 1). Marantz puts flash in the field. *Radio World,* pp. 41, 44.

Martin, D. (2000, July). 2000 salary survey. *BE Radio,* pp. 34, 44, 46, 48, 50.

Maunez-Cuadra, J. M. S. (1995, Summer). WUCF-FM's good-bye to carts: Our minidisc experience. *Feedback, 36*(3), 19-22.

Maycock, R. (2000, November). Choosing a digital audio workstation: Considerations for getting it right the first time. *Mix,* pp. 64–66, 68.

McGinley, T. R. (2001). A quarter-century of technical change. *Radio World silver source book 2001.* Falls Church, VA: Author.

Mims, B. W. (1998). Radio technology. In M. A. Blanchard (Ed.), *History of mass media in the United States: An encyclopedia* (pp. 570–574). Chicago: Fitzroy Dearborn.

Mitchell, J. (2000, January 3). E-mail correspondence with author.

O'Donnell, L. B., Hausman, C., & Benoit, P. (1989). *Radio station operations: Management and employee perspectives.* Belmont, CA: Wadsworth.

Otari (1989, October 11). Radio's robust "workhorses." *Radio World,* p. 13.

Pizzi, S. (1988, September 1). The state of the art of digital technology. *Radio World,* pp. 30, 38.

Rau, M. (1989). Radio's crystal ball. *The Radio World annual.* Falls Church, VA: Author.

Reist, N. (1993). PC revolution ups the education ante. *Radio World 1993.* Falls Church, VA: Author.

Rowland, S. (1994, March 23). DAWs make mixing documentaries easier. *Radio World,* pp. 17, 25.

Smith, F. L., Wright II, J. W., & Ostroff, D. H. (1998). *Perspectives on radio and television: Telecommunication in the United States* (4th ed.). Mahwah, NJ: Lawrence Erlbaum Associates, Publishers.

Tucker, R. K., Weaver II, R. L., & Berryman-Fink, C. (1981). *Research in speech communication.* Englewood Cliffs, NJ: Prentice Hall.

Wilson, B. G., & Dudt, K. P. (2001). College radios at Pennsylvania state universities: A status report. *College Media Review, 39*(1), 4–9.

Yu, J., & Cooper, H. (1983, February). A quantitative review of research design effects on response rates to questionnaires. *Journal of Marketing Research,* 20(1), 36–44.

Zarecki, T. (2001, April). *The art of voice tracking.* Presentation at the convention of the National Association of Broadcasters, Las Vegas.

CHAPTER 12

Learning Beyond the Classroom: Student Stations, Internships, and Distance Education

Kevin C. Lee

> *Too often, I fear, we have reacted to technology, rather than thinking creatively about how it might enrich our basic educational mission.*
>
> —*Stanley Katz* (2001, p. 1)

The incorporation of newer technologies into teaching—in broadcasting as in other disciplines—is a process of addition rather than subtraction or substitution. It adds to the cost and complexity of broadcast programs. It adds to the time demands placed upon educators and thereby may prove daunting to some faculty. However, the introduction of digital broadcasting to the instructional mix also adds a number of dimensions to pedagogy, including the expansion of learning and skill development beyond the classroom.

The continuing infusion of technology into broadcast curricula, though, is more than just something extra. In a number of dimensions it changes the educational process for everyone involved. This chapter focuses on the interaction of broadcast technology and learning in a variety of contexts in which students interact with learning environments and educators—formal and informal—outside standard class time. It also addresses some of the possible implications for institutions, instructors, and learners inherent in the digital delivery of video/audio content for the purposes of distance education.

Broadcast faculty who must familiarize themselves with "new media" applications regularly face hardware and software with steep learning curves. Multiplied hours must be invested at the outset for faculty to become conversant with the increasing complexity of new machines, operating systems, and software. Tenure-track educators wonder aloud how these time investments will benefit their cases before promotion and tenure committees. And before incorporating advanced gadgetry into their familiar course structures, instructors seek to gain a measure of their effectiveness and appropriateness in the domain of a focused area of broadcast study.

Faculty in broadcast programs face the reality of having to entertain a variety of new disruptions to the teaching-learning process: design and refine assignments utilizing digital technology, troubleshoot problems with hardware, software, and content delivery systems, and direct student acquisition of skills necessary for interacting with the various components in new media systems. Additionally, they must develop and/or redesign measures for assessing student work. All these factors add complexity to faculty responsibility as students strive to complete assignments outside the classroom.

Digital broadcast systems also are expensive to acquire and maintain, and quickly become dated. In an era of cost-conscious institutions, broadcast faculty often are drawn into departmental and campus-wide debates regarding the necessity of specific equipment expenditures (Barner, 1996).

Despite the formidable challenges associated with new audio and video digital systems, advanced technology in the broadcast world is not going to disappear or even diminish. From the smallest rural stations to the largest networks, audio and video professionals already have surrendered their facilities to a swift, irreversible course of digital creation, delivery, and retrieval systems. The New York City market, for example, has its own vibrant New Media Association, and estimates put the number of multimedia jobs there at the turn of the millennium into six digits (Ianzito, 1996). Television stations in even the smallest markets are now using digital three-chip cameras for ENG, nonlinear systems for editing, and taking steps toward the FCC-mandated deadline for broadcasting digital content for HDTV. Radio stations feature music in a variety of digital formats, programming delivered via satellite, and on-site computer systems monitoring the signals continuously. Broadcast educators, seemingly, have little choice but to immerse their students in the digital stream, both in the classroom and beyond.

However, the acquisition and incorporation of technology into the educational process stirs at the very center of the decades-long debate regarding journalism and broadcast curricula: Will the student be served best by an approach that emphasizes a broad liberal arts foundation, or by an approach that emphasizes pragmatic career preparation, or by one that stakes out a middle ground somewhere between programs for the Renaissance man and the polished preprofessional? (Kock, Kang, & Allen, 1999; Hilt & Lipschultz, 1996; Rosenbaum, 1992; Christ, 1990)

The liberal model builds on the understanding that college-educated persons should be equipped to access information from a plethora of sources, become critically conversant with it, and communicate clearly about it. The model also responds to outside criticism that broadcast programs offer little more than trade schools:

> Being responsive to student demands for a marketable skill upon graduation is shortsighted. In the true final analysis of the role of vocational education vs. higher education, entry-level needs of employers are served rather than educating for a life-long career and professional growth (Moore, 1992, p. 16).

In short, college-educated broadcast professionals should cultivate the habits of lifelong learners: critically reading and reflecting and responding. This is a perpetual challenge and requires a concerted effort on the part of the educators and students.

Hilt and Lipschultz (1996) believe that a "proper mix" of classical and concrete learning is feasible:

> By implementing changes that will enhance hands-on experiences without diluting the liberal arts curriculum, we are being more responsive to our students' needs. A student who comes out of college with a broad liberal arts education, strong communication skills, a desire to succeed, and some hands-on industry internship training has an advantage in landing a job in a highly competitive field (p. 42).

When it comes to curriculum, both the professional model and the blended liberal-professional model include a substantive experiential component. This acknowledgment of the need for a real-world connection is consistent with feedback to broadcast schools from outside organizations (Roper, 1987).

In practical terms, student learning is "the acquisition by students of any lasting knowledge or skill consistent with the educational purposes of the institution" (Kuh, 1991, p. 6). In this broad sense, pragmatic student learning clearly is taking place routinely both inside and outside of scheduled class time.

Astin (1985) argued that the most excellent education occurs in those venues that "add the most value" to students' scholarly and personal development. Educators today expend a great deal of energy developing and revising curricula, classrooms, and libraries, and rightly so. These are crucial components at the epicenter of dynamic learning. Beyond those perennial venues, a significant amount of crucial student learning occurs in those often unofficial experiences that take place regularly outside the confines of the traditional classroom.

> Out-of-class environments, however, are rarely neutral with regard to student learning. It seems reasonable to assume that when out-of-class experiences complement the institution's educational purposes, they contribute significantly to student learning and personal development (Kuh, Schuh, Whitt, & Associates, 1991, p. 6).

One of the most direct positive correlations between in-class and out-of-class modes involves reinforcement of disciplinary constructs. Gardner and Van de Veer (1998) state "student experiences outside the classroom often serve as vehicles for practice and application of course material" (p. 9). For instance, the broadcast student creating a video for promotion of his or her athletic team or serving as a DJ at a friend's wedding reception can be viewed as engaged in practical learning that provides a seamless connection with disciplinary course content.

The traditional delineation between curricular and extracurricular enterprise is determined by the awarding of academic credit toward a degree. Out-of-class experiences, therefore, include those for which academic credit cannot be earned.

Kuh et al. (1991) outline a number of different experiences in which out-of-class learning may take place: interaction with faculty outside of class, research projects, teaching responsibilities, orientation programs, employment (on- and off-campus), internships, community service, and activities with clubs, student organizations, student residences, recreational sports, and athletics (intercollegiate and intramural).

Several studies (Kuh et al., 1991) indicate that a number of positives accrue for students involved in out-of-class activities compared with their fellow students who were not involved extracurricularly:

- more likely to attribute some of their job success after college to out-of-class activities
- more likely to graduate
- more satisfied with their academic major
- more satisfied with contacts with faculty
- more satisfied with their living environment
- more satisfied with their social life

Gardner and Van de Veer (1998) conclude: "Institutions should be more intentional about linking in-class and out-of-class experiences, thereby encouraging higher levels of student learning (Kuh, 1996)" (p. 9). Recommendations for extracurricular learning, though, might best be viewed in light of the growing number of nontraditional students, many of whose lives often revolve around the beyond-the-campus world. For the nonresidential commuter, the student with a full- or part-time job elsewhere, applying course concepts often involves routinely taking classroom knowledge and skills far beyond the borders of the university into the realms of work, home, church, and community.

Fisher, Dwyer, & Yocam (1996) argue for a move from the more traditional "knowledge instruction" to a "knowledge construction" (p. 20) learning environment. Knowledge instruction includes didactic, instructor-centered activities with the faculty acting primarily as expert "fact tellers" and the students acting as learners in the "listening" mode, accumulating facts and using technology in a drill-and-practice setting. Transitioning to knowledge construction means migrating toward more learner-centered activities with instructors serving as collaborators (sometimes learner, sometimes expert), exploring the relationships among concepts through inquiry and invention, using technology to communicate messages through collaborative projects, expression of ideas, and data access.

Many exercises involving broadcast technology outside of traditional courses are practical in nature and may readily incorporate knowledge construction—from recording actualities for a radio news program to editing video of a dramatic presentation using nonlinear digital equipment. Raizen (1995) notes that students benefit from such pedagogical strategies because they are "learning through practical engagement in investigating, designing, making, testing, or improving a variety of objects, systems or environments" (p. 53).

Because many exercises involving new technologies can be inherently entertaining, the pedagogical challenge is to ensure that physiologically and emotionally engaging activities also are cognitively stimulating and enriching; hands-on tasks also should be "minds-on" assignments. "Such activities must elicit levels of critical and creative thinking that help students question, rethink, and reconstruct their understandings" (Raizen, 1995, p. 53).

Instructors creating assignments for broadcast learners operating beyond the physical environs of the classroom may build them in a progression. The following typology, for example, is designed to move students incrementally to higher levels of learning (Raizen, 1995):

1. Assignments that help build foundational knowledge and skills that will be used throughout students' experience in the major and in their careers
2. Tasks through which the standard procedures and professional practices for using technologies in the discipline are introduced and reinforced
3. Projects for which learners are required to apply problem-solving strategies in order to successfully negotiate
4. Case studies for which students are required to analyze and evaluate a broadcast event in its particular context—the analysis dealing with the influences leading to the event, the evaluation dealing with the effects on individuals and society

Technology enables broadcast educators to expand learning beyond the classroom as they direct students to venues where they interact with technology at their own pace and skill level. With thoughtful design, judicious feedback, and focused assessment, practical assignments can be greatly enriching for broadcast students.

DIGITAL CONVERGENCE AND STUDENT BROADCAST STATIONS

Faculty news-seekers along a hallway of academic offices in January 2001 consulted a variety of media sources as they awaited the U.S. Supreme Court's decision on disputed presidential election results in Florida. Fittingly, perhaps, the first report of their decision came from a digital version of one of the earlier of mass media: *New York Times* online. Radio and television reports rapidly followed.

Students entering a broadcast career from college must be prepared for a mediated world that is characterized by a dizzying spectrum of digital convergence. News sources on the World Wide Web provide provocative examples: online newspapers, such as washingtonpost.com, regularly include video and audio clips of important news events. Their competition includes more than other digital papers: online sites from television sources—such as CNN, ABC, and even local stations—include detailed textual stories to accompany their broadcast sound bites. Information seekers can listen to the day's version of news from National Public Radio at any time over the Internet, or search its archives and play back stories from last week, last month, or last year.

The FCC began to move radio in the direction of digital signals in 1992 with the approval of digital audio broadcasting (DAB) for satellite and antennae transmissions (Dizard, 1994). Stations initially have been reluctant to make the transition to DAB because of the high cost of replacing their equipment, the cost to consumers for replacing their many radios, and competition from cable and Internet content providers. Similarly, over-the-air television stations have been plodding along in the expensive transition to high-definition television (HDTV), and viewers have balked at high-priced new sets and picture quality not significantly better than many existing sets. Cable channels appear to be a step ahead of traditional broadcasters in the move to television that is all-digital all the time.

Despite the loss of Napster's status as the online source for unlimited free songs, "almost-on-demand" Internet interactive radio sites—which allow listeners to select genre of music and frequency of artists—are providing a similar service to Web audiences. In May 2001, popular services MTV Launch, MTV SonicNet, and MusicMatch all attracted over 1 million listeners, and over 75 million persons in the United States say they listen to Net radio regularly (Graham, 2001b). A growing number of persons are recording their own CDs at home from MP3 files and plugging them into new car stereos equipped to play downloaded files (Graham, 2001a). Up to 150 of the smaller audio files will fit on one disc.

The National Association of Theater Owners soon expects electronic motion picture projectors from JVC-Hughes and Texas Instruments to begin replacing the antiquated persistence-of-vision celluloid machines that have held reign in the film world for nearly a century (Sterngold,

1999). Instead of bouncing around in the back of a FedEx truck, movies soon could be sent to theaters by satellite and stored on a computer server. Mainstream films then would take their place alongside the digital short films now available on the Internet from the current Big Three of online video entertainment: Shockwave, iFilm and Mondo Media (Graham, 2001c).

Arbitron and Nielsen have been experimenting with Personal People Meters (PPMs) to keep track of the dizzying array of consumptive choices (Epstein, 2001). PPMs are worn by randomly selected participants during waking hours, silently make a record of every radio, TV, or streaming-video signal near it and—while recharging at night—send the data to the survey company.

Today individuals can turn on their computers to listen to the music they preselect, record their own CDs for car or home, read/listen/watch live or archived news 24/7, view film shorts, and communicate to others around the world via text, voice and video. All the while, portable devices track their consumptive activity for the ratings race. This is the world broadcast graduates are entering now; the future is difficult to predict.

The watchword for tomorrow's digital news gatherers and broadcasters will be the same as it has been: get more information and deliver it faster. What has changed is the technology of sending and receiving. Computer-assisted reporting has become "part of the standard repertoire in many of the nation's newspapers" (Lee & Fleming, 1995, p. 23). And the race to be the first with "breaking news" has been turned on its head as online versions of print newspapers can update their sites instantaneously, competing with radio and CNN for spontaneous news updates. Unless it is an urgent matter, the rest of television will wait for scheduled breaks before announcing news that "broke" on the Net, cable, or radio. However, with a news race comes risk, notes Leonard Shyles (personal communication, May 17, 2001)—Associate Professor of Communication at Villanova University and author of *Video Production Handbook* (1998):

> There will be increased pressure on the editorial process to verify news influenced by just-in-time material. Snap decisions will become more commonplace, news checking will unfortunately suffer, and we're back at square one, with respect to the credibility issue again.

Additionally, the career world beyond the university now includes venues and job titles and responsibilities that didn't exist even a decade ago. After graduating from Columbia University Graduate School of Journalism, Robin Sparkman landed a job as an "online producer," instead of moving into a traditional role in a print or broadcast newsroom (Ianzito, 1996). Her cutting-edge position involves translating video and audio from CNBC-TV programs to digital form for the Web.

In a field where clear delineations between print, radio, and television have existed for decades, broadcast programs may benefit from some rethinking as they help students develop a metaview of a new mediated world order. The industry will require future practitioners to move fluidly among the plethora of technological changes, perhaps requiring them to frequently transition among tasks and responsibilities.

A number of educators believe the student-run broadcast station is an ideal place for experience and experimentation with new media. Many students may have opportunities to work with digital technologies at campus radio and television stations or independent production projects for campus entities. The intended audiences for these venues are often other students and faculty who are assessing student work.

Damian Little (personal communication, June 15, 2001), a senior at a private college in the Mid Atlantic and general manager of the campus station, has found his work with student radio to be highly beneficial for learning about audio technology. He moved the station's production from analog to digital using mini disk and CD players, a studio computer, and Cool Edit for editing .WAV format sound files. This system enabled his DJs to record air checks, prerecord their shows, "modify" obscene lyrics, and add effects—such as echo and reverb. Commercials and PSAs also were prerecorded and edited on the computer before airing. "The computer has saved us time," Little said. "Making a 60-second commercial in the old days would take up to 2 hours; now with the computer and digital editing it is like 10 to 15 minutes to make."

With help from another student, Little wrote a software plug-in for Winamp that would run the station by itself. "To get software made for audio broadcasting costs way too much," Little noted. The general manager also has used a program called VNC (Virtual Network Computing) developed by AT&T labs in England. VNC is a remote display system that allows users to view a computing desktop environment not only on the machine where it is running, but also from anywhere on the Internet and from a wide variety of machine architectures (VNC, 2001). "This software lets us run the station computer from any computer on campus," he explained. "We use this to write messages to the DJ, or we can start a song from any computer if the DJ has dead air." The senior's experience with the radio station also has included assisting with the introductory radio course, supervising executive and sales staffs, participating in a national conference, and putting the station on the Web (Stephens, 2001).

There are many broadcast students who may have had opportunities for similar experiences with technology—working to digitize radio broadcasts and operate student station Web sites. McClung (2001) notes that the audience for such broadcasts is diverse and "sizeable" (p. 72).

Clearly, experience with campus stations can be quite beneficial for development of technological skills for a number of students, but in a number of ways it cannot fully imitate a professional setting. Holton (1998) states:

> The work world is so fundamentally different from the world of education that it requires an almost total transformation on the part of the new graduate . . . Many of the skills students developed to be successful in education processes and the behaviors for which they were rewarded are not the ones they need to be successful at work (pp. 100–101).

In an attempt to more closely approximate the professional newsroom of the future, Brigham Young University merged its traditional daily paper and broadcast entities in early 1996. In their place, they implemented NewsNet, a converged student-run entity producing news products for print, radio, television, *and* Web. The heart of the system was Daybook, a database of stories that reporters were covering (Hammond, Petersen, & Thomsen, 2000). According to the student online editor, "The Daybook picks up anything new, no matter what area. [It's the] brink of something totally new, . . . information at [your] fingertips. All reporters have the responsibility to update the stories they are preparing in the Daybook" (Hammond et al., 2000, pp. 21–22).

The good news: *Editor and Publisher* recognized NewsNet as the highest ranked student online service in both 1998 and 1999. "The combination of visual and textual material has been cited as a major reason" (Hammond et al., p. 24). New job roles were developed in the converged newsroom, and students who interned at NewsNet have found that they are better prepared for the job market.

> In the old organization, students were encouraged to specialize on certain media, on certain beats, or on certain skills such as copy editing or videotape editing. In the new organization . . . students were more highly valued if they had integrative skills that allowed them to contribute across media (Hammond et al., p. 21).

However, for the BYU station, the challenges for creating and operating the multimedia operation have been daunting (Hammond et al., 2000):

- Substantial investments of time and resources were required at the outset
- Many students enrolled in programs had envisioned themselves in medium-specific roles: television anchor, radio announcer, print reporter
- Traditional delineations between print, radio, and television were reinforced by curricular structure, course conten, and instructor orientation
- The converged newsroom operated much differently than the traditional organization, less top-down command flow and defined job tasks, more lateral communication and negotiation

The researchers noted that ". . . this self-organizing, seemingly chaotic work environment where processes were continually improvised still needed to meet deadlines, produce shows and get the paper to press. Medium based jealousies persisted" (Hammond et al., 2000, p. 21).

As harried and competitive as many student stations can be under deadline pressure, they still cannot fully re-create the intensity of a professional news enterprise. Ehrlich (1993) observed and interviewed television reporters, producers, and supervisors in both a medium-sized and top-20 market for six months, arriving in the early afternoon and staying through the late news broadcast. He describes a culture in commercial stations where "competition is 'ritualized' and built into the norms and routines. Newsworkers strategically and ritualistically monitor other stations in order to reassure themselves that they are not getting beaten on stories" (pp. 201–202). They also often made "derogatory remarks about the other stations' anchor teams and stories" as they watched during breaks and "routinely monitored what rival stations' news crews were doing by quizzing reporters" (p. 200) returning from the scene of news events.

Perhaps more constructively, there are time-honored criteria for decision making and on-air presentation that permeate most professional stations. These cannot be ignored in the race to become proficient with new media applications:

> As a videographer, there are today—as in the past—techniques of presenting news stories clearly and compellingly, not to mention accurately. These goals transcend the newest technological breakthroughs. They represent the enduring goals of good journalism (Leonard Shyles, personal communication, May 17, 2001).

INTERNSHIPS

Barner's (1996) survey of 200 broadcast graduates from BEA member schools who were working in the industry found that 81% strongly favor a required full-time internship at a broadcast station as part of the curriculum. Echoing the observations of Holton, noted above, the respondents expressed the opinion that "the campus station could not adequately simulate the commercial broadcasting environment" (p. 24). Additionally,

> Written responses show that they [survey respondents] feel the internship . . . gives the student *a chance to experience new technologies* that are not affordable in many colleges (p. 24, emphasis added).

The opportunity to "experience new technologies" itself was an important reason for requiring internships, Barner's respondents indicated. Because of cost considerations and the methodical pace of change in programs and institutions of higher education, broadcast programs lag behind commercial stations in the acquisition and utilization of digital equipment.

Hilt and Lipschultz (1996) describe the internship as a kind of bridge between college and career. They are "an extension of journalism and broadcasting courses and an important part of the curriculum (Hilt, 1991)" (p. 37). Over 96% of their student survey respondents—junior and senior interns during their final week in the cooperative program— "agreed that internships prepare future broadcasters and journalists" (p. 39).

Eyler (1995) details the beneficial blend of intellectual and professional learning that field experiences, ideally, can provide:

> Many of the goals articulated for college internships are quite consistent with those identified for both a liberal education and experiential education. These include putting theory into practice, acquiring a deeper understanding of how organizations operate, developing higher order cognitive skills, becoming independent learners, increasing personal growth, developing mature values, becoming responsible citizens, and exploring careers and acquiring career competence (Couto & Zuberer, 1988; Stanton, 1990) (p. 186).

Internships for university students come in many shapes and sizes. Although they may or may not include pay, faculty supervision, academic credit, or be completed during the academic year, they most often include learning that falls outside the rubric of traditional in-class learning. In broadcast programs, internships have become an integral component of students' educational experiences; in 95 percent of the programs, they are part of the curriculum (Hilt & Lipshultz, 1996; Basow & Byrne, 1993; Meeske, 1988).

There are several types of internship programs, including alternating—where students spend a full semester, five days a week, at field sites; parallel—where students are completing courses and working at field sites concurrently; and independent study—where students work as volunteers at field sites under the purview of an individual faculty member, completing papers and projects for academic credit, usually from one to four hours (Hanson, 1984).

Often, students must meet certain criteria to be eligible for an internship. These may include junior or senior standing, the successful completion of a particular number of hours in the department and/or certain prerequisite courses, the maintenance of a minimum GPA (in departmental courses and/or in all university coursework), approval of their departmental adviser or the program's internship course supervisor/committee (Watson, 1992).

Some programs also require interns to complete an internship course concurrently with the field experience. Such courses may include assistance with application materials—such as resumes and audition tapes, interview preparation, compilation of a daily journal of experiences, written critiques of journal articles relating to the internship, periodic oral reports of internship experiences, selected readings, case study analyses, research papers, and meetings with station supervisors (Watson, 1992).

Technological applications in the internship course could begin with the process of translating application materials into digital form and continue with the critiquing of articles relating to the use of technology in the contemporary newsroom or broadcast station. Regular correspondence among intern, professional supervisor, and internship course director may take place online via e-mail or postings to a Web site.

Internships provide opportunities for students to interact with new technologies in the context of a cohort of broadcast professionals for a "real" audience, and in a competitive environment. Shyles (personal communication, May 17, 2001) argues that all students who want to work in video journalism should have internship experience. He notes that internships expose students to "the folkways and practices of different organizations, and that range of experience can only be acquired in an internship. It is invaluable." In preparation for the demands of a broadcast career, he notes: "Daily interaction with the job and equipment is a huge help."

The transition from educational to professional worlds includes the necessity of becoming conversant and proficient with new technologies in order to be productive. Ciofalo (1989) notes: "Both teachers and bosses sit at the top of a hierarchical learning pyramid in which students and workers must assimilate a fixed body of knowledge to pass tests and complete projects" (p. 28). One of the more recent studies of internships found that broadcast graduates particularly may indeed need a boost in the form of practical experience as they face a challenging job market (Beard & Morton, 1999). "College graduates are increasingly competing with the underemployed and victims of corporate layoffs, who already possess substantial work skills" (p. 45).

One junior intern (personal communication, June 15, 2001) noted that the small-market station at which he was working featured predominantly satellite-fed programming run by a computer called Phantom, used by a number of stations around the country: "Getting to know how to run it was helpful. . . . This computer has the ability to edit sound files and work with on-air production." The intern added that his experience moved beyond that of a typical student station: "This computer is very pricey. Schools cannot afford this; very few college stations have this system."

An internship experience can help students move beyond the routine of course work if the self-reflective domain of personal growth becomes more engaged—the affective process that "envisions a future self" (Mentkowski & Associates, 2000, p. 186). In these settings, student interns may come to an understanding, for instance, that they would rather work in production than on-air, as they initially had dreamed. They also may be frankly confronted with the query, "Is this really what I want to do with my life?"

Additionally, broadcast students working at a professional station may be prompted to move beyond the initial thrill of simply using top-dollar digital equipment to the desire to produce the highest quality broadcast product for their audience, regardless of the technological sophistication. Educators hope interns also will make some applications of theoretical constructs as they immerse themselves in a field internship.

Tommerup (1993), as cited by Smith (1998), notes:

> Students often comment that internships help them see that their undergraduate work has indeed prepared them for the world of work. As one senior student noted: "They talk about combining theory with practice. Well, that is exactly what an [internship] does. But it also familiarizes the student with the demands of the working world. So, while you're still in school, you realize what kinds of skills you're going to need when you graduate" (pp. 90–91).

Based on their research with broadcast interns, Basow and Byrne (1993) recommend that internship supervisors

- Assist interns in the process of developing learning goals prior to beginning their fieldwork the students are making contributions to the station and should reasonably expect some benefits
- Assist interns as they form expectations prior to their field experience discussing the experiences of past interns with them may be particularly helpful
- Encourage interns to demonstrate initiative students may benefit from the knowledge that it may be helpful to ask questions and volunteer for appropriate duties at a busy station
- Encourage interns to consider multiple internships before they graduate perspective and opportunity may be enhanced by follow-up field experiences
- Encourage interns to make intellectual connections with their field experiences afterward syntheses of theory and practice can be made after students have finished an internship

Gray (2001) agrees that internships benefit from a clear articulation of the vision for the field experience. She also advises that steps should be taken to integrate the intern into the workplace; that the responsibilities of the intern be gradually increased; that sessions for evaluation of the intern's performance be scheduled regularly; and that the university supervisor be apprised of problems—whether from culture, climate, or conflict—when they occur. "One can read and study all the books and journals available; however, the ultimate test comes when the individual is actually in the trenches doing what must be done" (p. 665).

Problems that sometimes occur with internships include the uneven quality of field experiences among interns in the same broadcast department, the placement of some interns in low-tier positions that involve little more than clerical tasks, conflicts with labor unions, and the paucity of detailed, constructive evaluations from station supervisors (Meeske, 1988).

Any of these can prove to be problematic for students seeking to enhance their technological skills in a field setting. If some interns in a broadcast department are given responsibility working with advanced equipment and others are stalled in low-tech posts completing routine tasks, the progress of the program and the students will be frustrated. However, from a meta perspective, there are a number of benefits for the plurality of interns to be gleaned from a quality field experience (Watson, 1992):

> The internship is designed to help students build a relationship with one or more professionals in the business world, make difficult career choices, experience an organization without lengthy obligation, and discover long-term interests and aspirations. The internship often provides a valuable work experience, a letter of recommendation, a personal reference, and, sometimes, a job offer (p. 437).

ONLINE CLASSES

Not only is technology changing pedagogy in the discipline of broadcast, but digital delivery of university courses and programs utilizing video and audio Internet content also is ushering in a new era of higher education: the push for online distance education.

The State of Pennsylvania has a department devoted entirely to distance education and defines the practice as "not simply the addition of technology to instruction; instead it uses technology to make possible new approaches to the teaching-learning process" (Gilbert, 2001, p. xvi). More simply, "distance education is learning without the physical presence of the instructor" (Mantyla, 1999, p. 4).

Distance learning—along with associated terms like extended education and open learning—is becoming increasingly associated with Web-based courses in some circles. However, it traditionally has included a wide range of delivery venues. Derek Roundtree (1998), Professor of Educational Development at the Open University, details the modes that have been used. In the print realm, these include books, periodicals and journals, study guides, self-teaching texts (such as tutorials in print), workbooks that accompany audiovisual material, and static visuals (photographs, maps, charts). Many of these can be readily converted into digital form and posted on Web sites. Audiovisual materials—such as audiocassettes, videotapes, and even broadcasts—also can be distributed to learners via the Internet. Now a traditional third leg of the distance education structure, human interaction from a distance, is feasible online via text (e-mail, interactive Web sites), voice, and desktop videoconferencing (Roundtree, 1998).

Distance learning, however, has not enjoyed a prestigious place in academe. According to *The Survey of Distance Learning Programs in Higher Education*, a report of the Primary Research Group (1997), until recently, education from a distance "enjoyed a reputation in the academic pecking order somewhat above mud wrestling and roller derby and somewhat below traditional academic education" (p. 13). However, digital delivery has brought dramatic changes:

> The astounding growth of the Internet, and the gradual though steady increase in the personal computer usage. . . . the deregulation of telecommunications markets, and the surge of new telecom technologies . . . helped shake the mud off of distance education and enshrine it as a technologically advanced, cool and academically sexy thing to be involved with (p. 13).

Without a steady stream of student enrollees, online courses and programs would eventually disappear. Therefore, the altering of the public's general perception of distance learning may be among the most significant accomplishments associated with putting courses online (Porter, 1997).

One recent video-intensive project involved students from Cornell and universities in India, Costa Rica, the Netherlands, Sweden, Australia, and Honduras. Students form discussion groups of 8 to 10 persons that include representatives from each institution. During 90-minute teleconferences every two weeks, video cameras sent live pictures to television monitors divided into quadrants at each location. At other times, students at institutions across 16 time zones dialogued via computers equipped with video cameras and microphones and NetMeeting software (Ludwig, 2001).

There are more than 100,000 distance courses offered among the 4,000 U.S. institutions of higher education, and the number is growing rapidly (Gilbert, 2001). Statewide organizations and virtual universities have been initiated in at least 35 states, according to the Instructional Telecommunications Council's (ITC) master list (Young, 2001). "Successful use of distance education in Canada, Australia and the UK has also led many American colleges (that already compete in world education markets with these countries for foreign students) to take another look at distance education" (Primary Research Group, p. 13). ITC Executive Director Christine Dalziel noted, "A lot of colleges are really rushing to put courses together, and they're very afraid of the competition from other states. That's really what's driving it in a lot of cases: fear" (Young, 2001, p. 1).

Additionally, a number of new entities have entered the foray—including some for-profit dot-com universities. Porter (1997) notes that "the role of traditional academic institution is changing; colleges and universities will have to compete with a growing number of other educational providers or vendors" (p. 19). The list of new institutions includes the University of Phoenix, Universitas 21 (linked to Rupert Murdoch's News International), Western Governors University, the National Technical University, World Community College, and Unext. Not bound by the constraints of political and geographic boundaries, online learning is becoming multinational, multidisciplinary, and multipurpose: serving both traditional and nontraditional students wherever and whenever they choose to log on and learn.

The League for Innovation in the Community College has been developing guidelines for its 750 member institutions to negotiate agreements with universities offering distance education courses, enabling community college students to readily transfer lower-division credits to institutions offering bachelor's degree programs online. Discussions are under way with Florida State University, the University of Maryland University College, and Capella University; deals have already been reached with Western Governors University, the University of Phoenix, and the United States Open University (Carnevale, 2001). Interestingly, the earliest agreements were signed with distance-only online institutions; the latter round includes traditional universities that are branching out.

Distance education is promoted by many as a means of bridging the gap between instructor and learner when separated by space, particularly, or by time. Some university administrators ultimately see it as a means to cut costs, increase enrollment, and increase revenue (Primary Research Group, 1997). Faculty perceive inherent threats to employment opportunities, compensation, job security and legal ownership of the pedagogical process and instructor-created content. In keeping with these concerns, many faculty groups are taking a "hard line" toward distance education ("Faculty union," 2001). Responding to numerous requests for information, ITC surveyed its membership regarding faculty compensation for distance education (Edwards & Minich, 1998). The researchers concluded broadly that institutions "need to develop incentive, reward, and compensation infrastructure that is equitable, cost effective and insures the highest possible quality" (p. 2).

The recently passed Technology, Education, and Copyright Harmonization Act enlarged the so-called safe harbor for copyrighted digital music and video works used in online instruction (Foster, 2001). This legislation may well serve to greatly expand the use of video and audio material in online courses and programs. However, Stanley Katz (2001), director of a policy studies center at Princeton, warns, "It is not clear that the traditional 'fair use' exceptions to intellectual property rights for scholarship and teaching will be protected under the emerging e-copyright regime" (p. 6).

Institutions anxious to wedge into this emerging market often partner with private companies using prepackaged Web delivery systems such as Blackboard, eCollege.com, Web Course in a Box, and ed2go.com. The one-size-fits-all approach, though, has some calling it "canned education" and questioning the quality of the instruction. Additionally, "the providers are usually anonymous and may or may not be connected with the college" (Gilbert, 2001, p. 48) that is providing the course.

Katz (2001)—stating that "too much of what is now being called distance education at most institutions is not an educational idea; it is a business idea" (p. 3)—wonders about the long-term implications of commercial-educational hybrid ventures:

> What is the difference between Yahoo or America Online and Columbia University? Less and less. The virtual commercial efforts of an increasing number of universities are evidence that the tool has become the goal. The larger question is whether the nonprofit university is subverting its mission in a quest for income from information technology. Does it matter that some of our finest universities are beginning to behave like proprietary educational institutions? Will entering the commercial sector eventually jeopardize their dedication to providing the highest-quality education they can? (p. 4).

Textbook publishers and suppliers of instructional materials recently have been touting the benefits of "using media to teach media" by supplementing traditional courses with audiovisual and Web-based resources. But larger questions loom on the horizon with the acceleration of online learning: Will broadcast courses, and even degree programs, be offered through distance education? Can broadcast courses be taught properly when they are largely, or wholly, taught online?

A cursory review of bachelor's degree programs available online in July 2001 uncovered no universities offering broadcasting degrees via distance education (Petersons.com, 2001). However—as the University Continuing Education Association (2001) points out—it is difficult in this rapidly expanding arena to conduct an exhaustive search of institutions, programs, and courses:

> Searchable course and program databases are proliferating on the Web. So far, none offers a complete search of all the distance education offerings of all the higher education institutions in the U.S. or worldwide (p. 5).

Some institutions offer individual media-related courses via the Internet, such as introductory surveys of mass communication. Pepperdine pilot-tested one unit of their four-unit course, Philosophy and Effects of Mass Communication, during fall 1999 (Rosenkrans, 2001). An evaluation survey was administered at the conclusion of the course and a summative committee created to review the results. Such methodical implementation and assessment of online course content may serve as a model for other programs.

Today many Americans have ready access to affordable multimedia computers, digital video cameras, and Internet connectivity. Through observation, experimentation, and mentoring, some create their own sophisticated audio files and video projects and post them on the Web. In time, some university broadcast programs may take advantage of resources like these and offer numerous courses online, including production classes and even degree programs.

Online distance education has achieved a great deal of momentum. Shyles (personal communication, May 17, 2001) observes that the so-called University Without Walls movement apparently "will not be stopped." Katz (2001) concludes:

> The new technology has unleashed such creative, frequently entrepreneurial activity that is so expansive, pervasive, and difficult to manage that it has had an impact on some of our fundamental practices in teaching and scholarship. It will continue to do so, and it will drive us if we do not drive it. Have we established the mechanisms to monitor and evaluate those developments? (p. 4).

REFERENCES

Astin, A.W. (1985). *Achieving educational excellence: A critical assessment of priorities and practices in higher education.* San Francisco: Jossey-Bass.

Barner, D.L. (1996). A national sample of recent graduates address broadcast curriculum. *Feedback, 37*(2), 22–25.

Basow, R.B., & Byrne, M.V. (1993, Winter). Internship expectations and learning goals. *Journalism and Mass Communication Educator 47*(4), 48–54.

Beard, F., & Morton, L. (1999, Winter). Effects of internship predictors on successful field experience. *Journalism and Mass Communication Educator, 53*(4), 42–53.

Borich, G.D., & Tombari, M.L. (1995). *Educational psychology: A contemporary approach.* New York: Harper Collins.

Carnevale, D. (2001, June 5). League for Innovation and Online Institutions Create Credit-Transfer Guidelines. *The Chronicle of Higher Education* [Online] Retrieved June 7, 2001, from http://chronicle.com/free/2001/06/2001060501u.htm

Christ, W. (1990). The broadcast model curriculum project: Developing national models. *Paper presented at the 76th Speed Communication Association Convention,* Chicago, IL.

Ciofalo, A. (1989, Winter). Legitimacy of internships for academic credit remains controversial. *Journalism and Mass Communication Educator, 43*(4), 25–31.

Dizard, W., Jr. (1994). *Old media new media: Mass communications in the information age.* White Plains, NY: Longman.

Edwards, R.E., & Minich, E. (1998). *Faculty compensation and support issues in distance education.* Washington, DC: Instructional Telecommunications Council.

Ehrlich, M.C. (1995). The competitive ethos in television newswork. *Critical Studies in Mass Communication, 12*(2), 196–212.

Epstein, W. (2001, July 12). Ratings technology takes turn toward accuracy. *The [Colorado Springs] Gazette,* Life 6.

Eyler, J. (1995, June). Graduates' assessment of the impact of a full-time college internship on their personal and professional lives. *College Student Journal 29,* 186–194.

Faculty union in Nova Scotia takes hard line on distance learning. (2001, June 15). *The Chronicle of Higher Education 47*(40) [Online] Retrieved June 17, 2001, from http://chronicle.com/weekly/v47/i40/40a02903.htm

Fisher, C., Dwyer D.C., & Yocam, K. (Eds.). (1996). i San Francisco: Jossey-Bass.

Foster, A.L. (2001, May 18). Senate committee favors letting instructors use more digital works in online classes. *The Chronicle of Higher Education* [Online] Retrieved May 21, 2001, from http://chronicle.com/free/2001/05/2001051801u.htm

Gardner, J.N., & Van der Veer (1998). The emerging movement to strengthen the senior year experience. In Gardner, J.N., Van der Veer & Associates (Eds.), *The senior year experience: Facilitating integration, reflection, closure and transition.* San Francisco: Jossey-Bass.

Gilbert, S.D. (2001). *How to be a successful online student.* New York: McGraw-Hill.

Graham, J. (2001a, July 17). Bye, Net: MP3s hit open road: Music-filled hard drive can ease a hard drive. *USA Today,* 3D.

Graham, J. (2001b, July 3). Net radio tangos with the law. *USA Today,* 3D.

Graham, J. (2001c, February 12). Suspended animation: If Icebox can't make it here, who can? *USA Today,* 3D.

Gray, T.I. (2001, May). Principal internships: Five tips for a successful and rewarding experience. *Phi Delta Kappan,* 663–665.

Hammond, S.C., Petersen, D., & Thomsen, S. (2000, Summer). Print, broadcast and online convergence in the newsroom. *Journalism and Mass Communication Educator, 55*(2), 16–26.

Hanson, J. (1984, January). Internships and the individual: Suggestions for implementing (or improving) an internship program. *Communication Education 33,* 53-61.

Hilt, M.L., & Lipschultz, J.H. (1996, Spring). Broadcast Newsroom Hiring and Career Preparation. *Journalism and Mass Communication Educator, 51*(1), 36–43.

Holton, E.F. III (1998). Preparing students for life beyond the classroom. In Gardner, J.N., Van der Veer, & Associates (Eds.), *The senior year experience: Facilitating integration, reflection, closure and transition.* San Francisco: Jossey-Bass.

Ianzito, C. (1996, November/December) It's a job, but is it journalism? *CJR* [Online]. Retrieved July 8, 2001, from http://www.cjr.org/year/96/6/newmedia.asp

Katz, S.N. (2001). In information technology, don't mistake a tool for a goal. *The Chronicle of Higher Education, 47*(40). Retrieved June 20, 2001 from http://chronicle.com/weekly/W47/i40/40b00701.htm

Kock, E., Kang, J.G., and Allan, D.S. (1999). Broadcasting education, curricula in 1-year and 4-year colleges. *Journalism and Mass Communication Educator, 54*(2), 4–15.

Kuh, G.D., Schuh, J.H., Whitt, E.J., & Associates (1991). *Involving colleges: Successful approaches to fostering student learning and development outside the classroom.* San Francisco: Jossey-Bass.

Lee, K.C., & Fleming, C.A. (1995, Autumn). Problems of introducing courses in computer-assisted reporting. *Journalism and Mass Communication Educator, 49*(3), 23–34.

Ludwig, J. (2001, 28 March). A course with participants around the globe looks at environmental issues. *The Chronicle of Higher Education.* [Online]. Retrieved March 28, 2001, from http://chronicle.com/free/2001/03/2001032801u.htm

Mantyla, K. (1999). *Interactive distance learning exercises that really work!* Alexandria, VA: American Society for Training & Development.

McClung, S. (2001, Spring). College radio station Web sites: Perceptions of value and use. *Journalism and Mass Communication Educator, 56*(1), 62–73.

Meeske, M.D. (1988, Summer). Update: Broadcast intern programs and practices. *Journalism and Mass Communication Educator, 43*(2), 75–77.

Mentkowski, M., & Associates (2000). *Learning that lasts: Integrating learning, development, and performance in college and beyond.* San Francisco: Jossey-Bass.

Moore, R.C. (1992, Spring). Let's retain the undergraduate core. *Feedback, 33*(2), 15–16.

Petersons.com (2001). Retrieved July 9, 2001 from http://www.petersons.com/dlearn/

Porter, L. R. (1997). Creating the virtual classroom: Distance learning with the Internet. New York: John Wiley & Sons.

The Primary Research Group, Inc. (1997). The survey of distance learning programs in higher education. New York: Primary Research Group, Inc.

Raizen, S.A., Sellwood, P., Todd, R.D., & Vickers, M. (1995). *Technology education in the classroom.* San Francisco: Jossey-Bass.

Roper Organization (1987). *Electronic media career preparation and study.* New York: The Roper Organization.

Rosenbaum, G. (2002). Design considerations for an effective online environment. *Jounalismand Mass Communication Educator, 56*(2), 43–61.

Rosenkrans, G. (2001, Spring). Design considerations for an effective online environment. Journalism and Mass Communication Educator, 56(1), 43–61.

Roundtree, D. (1998). Exploring open and distance learning. London: Kogan Page Limited.

Shyles, Leonard (1998). *Video production handbook.* Boston: Houghton Mifflin.

Smith, B.L. (1998). Curricular structures for cumulative learning. In Gardner, J.N., Van der Veer, & Associates (Eds.), *The senior year experience: Facilitating integration, reflection, closure and transition.* San Francisco: Jossey-Bass.

Stephens, P. (2001, Spring). On the air 24/7: C-92 online and around the world. *The Orange Line,* 4–5.

Sterngold, J. (1999, February 22). A preview of coming attractions: Digital projectors could bring drastic changes to movie industry. *New York Times,* pp. C1–2.

University Continuing Education Association. (2001). Retrieved July 12, 2001 from nucea.edu/Distance2.htm
VNC. (2001). Virtual Network Computing. Retrieved July 12, 2001 from http://www.uk.research.att.com/vnc/
Watson, K.W. (1992, October). An integration of values: Teaching the internship course in a liberal arts environment. Communication Education 41, 429–39.
Young, J.R. (2001, 13 June). Web site tracks statewide virtual university projects. *The Chronicle of Higher Education.* [Online]. Retrieved June 15, 2001, from http://chronicle.com/free/2001/06/2001061301u.htm

CHAPTER 13

Characteristics of Broadcasting Curricula: Perspectives from Three Continents

Seok Kang and Jong G. Kang

Across the world, broadcast education has been impacted by technological changes (Blanchard & Christ, 1993). Chang (1997) presumes the structures and processes of broadcast education around the world "can reasonably be expected to follow the interplay among political, social, cultural, and economic factors within and outside a particular country's geographical domain" (p. 4). The outcomes of broadcast education are mediated by the structure of curricula developed in that country. This structure will influence future media practitioners who study the broadcasting major. As such it can positively or negatively shape the future of the media industry in that country. In this view, examining the broadcasting curricula in a country is a first step toward predicting the media environment of that country.

This study examines the broadcasting curricula at institutions of higher education in three countries: the United States, Britain, and Korea. It asks whether the broadcasting curricula primarily emphasizes the acquisition of practical skills or theoretical knowledge. It also asks to what degree these curricula are responsive to technological changes in the media industry.

Because issues such as the impact of technology on changes in higher education are global (Blanchard & Christ, 1993), research in this area should extend beyond the borders of the United States. Throughout the world, educators agree that institutions of higher learning must train students to think critically and to develop skills of analysis and interpretation. At the same time, many believe it is imperative that these institutions also meet the immediate needs of industry and the job market.

The researchers chose the United States, Britain, and Korea for comparison of broadcasting curricula for several reasons. First, this study included British institutions because the authors were interested in the curricula of the country of the BBC, which represents a model of public broadcasting. Second, there has been little research done on the broadcasting curricula of Britain. Third, the analysis of broadcasting curricula in a European country provides a suitable comparison with curricula in North America and Asia. The researchers chose Korea for comparison because they are both native Koreans who speak and write Korean and English. They have experienced the educational systems of both countries. Also, they have witnessed an increasing interest in broadcast education in Korea.

LITERATURE REVIEW

Theory or Practice

Historically, there have been two main sides in the discussion of curricula at broadcasting programs. One side emphasizes theoretical knowledge while the other side emphasizes the acquisition of practical skills. A third view seeks to honor both perspectives and asserts that theory and practice should be stressed together.

Researchers have documented these views. Forty years ago, Niven (1961) found that the three large classifications of broadcast curricula are Liberal, Practical, and Liberal-Professional. "Liberal" refers to theoretical, historical, or ethical subjects in mass communication courses. "Practical" includes TV production and radio production, which require practical skills. "Liberal-Professional" provides students with practical studies as well as theoretical-driven courses.

More recently, Blanchard and Christ (1993) contended that mass communication education usually emphasizes one of the three categories: "(1) courses that train students for entry-level positions in the trades (e.g., broadcast news, newspaper journalism, video and audio production), (2) courses that study the industry so that students can fit into the industry (e.g., communication law), and (3) courses that critique the industry so that students can be an informed public, informed consumer (e.g., mass media and society courses" (p. 85). These categories represent the typical formation of today's current mass communication curricula.

In early years, Pennybacker (1965) pointed to the conflict between the philosophy of a liberal arts education and that of training for a specific skill. For instance, some academic institutions posit that a liberal arts education is important because it provides students with "concepts, critical insight, and discussion and analysis of issues" (McCall, 1987, p. 19). A liberal arts education aims to not only provide knowledge but also to help students foster an ability to think critically and to evaluate social phenomena, thereby being a competent citizen in society. It is generally agreed that some background in a liberal arts education is critical, regardless of a student's major.

Meanwhile, as technology and the media market grow, practical training has also been emphasized. For example, Renz (1991) underscores the necessity of practical education in broadcast education. Renz states that "communication skills are an integral part of the art and industry of broadcasting; therefore, the general practice of teaching these skills should be retained as an integral part of the broadcast education program" (p. 22). Murray (1992) claims that the objectives of introductory courses to mass communication include two components: one general and philosophical in nature, the other specific and related to the acquisition of skills. In summary, broadcast curriculum generally consists of some mix of theoretical and practical courses. This model will be applied to our study of broadcasting curricula in the three countries.

Interplay of Broadcasting Curricula With Facilities

There appears to be a relation between media facilities and a nation's prevailing academic priorities in broadcasting education. For instance, in the United States, the broadcasting industry consists of both commercial and non-commercial broadcasting companies. In addition to nationwide networks, more private production companies and local stations exist in the United States than any other country in the world. The increased job opportunities due to this expansive industry has yielded the argument for a strong relationship between broadcast education and its industry.

The concern that broadcasting curricula need to be reassessed and reconsidered in view of technological changes is expressed in many studies and articles. Some arguments claim that broadcasting curricula in the United States is likely to be more practice-driven (BEA, 1990; Blanchard & Christ, 1993). Barner (1996) contends that students should be well prepared in practices in order to "function in a competitive environment" within a media system.

A rather conservative academic environment in Europe tends to affect the acceptance of any new discipline (Giglioli, 1994). In the case of Britain, conservative sentiments toward the broadcasting major have prevailed since at least the 1960s. Bannerman (1967) claimed that the educational philosophy of Britain has always emphasized preserving a strong academic respectability grounded in the liberal arts with a cautious approach taken toward any curricula that might be viewed as vocational.

This tradition, however, is currently being reexamined because of media developments in Europe (French & Richard, 1994). The changes observed with the new BBC charter due in 1996, and the joining of four new commercial television companies into a different ITV (Independent Television) system (Monush, 1995), indicates that those changes will continue to influence university curricula (French & Richard, 1994). Even though present broadcasting curricula in British institutions is more theory-oriented because practical sectors are mostly managed and run by broadcast firms (Morrison, 1997) and polytechnics, technological developments are influencing broadcasting curricula in Britain.

Asia is no exception when it comes to the impact of technology on broadcasting programs. For instance, in the early 1990s, Korea entered the era of multichannel systems with an increase in networks. In 1995, Korea experienced the advent of satellite broadcasting and the launching of cable television systems. This growing media industry has resulted in competition among networks (Kim, Kim, & Kang, 1994). As the broadcasting industry enters into a stage of rapid development, the job market in broadcasting calls for personnel that are both practically and theoretically qualified.

The change in Korea resulted in the establishment of in-house training institutes by broadcast organizations. Currently, "MBC (Munhwa Broadcasting Corporation) and SBS (Seoul Broadcasting System), both commercial television networks, run "MBC Broadcast Institute for Culture and Arts" and "SBS Broadcasting Academy." KBS (Korean Broadcasting System), a public-owned station, entrusts one university (Sogang University) with practical education for practitioners.

For this reason, university graduates who want to enter the broadcast job market are reeducated at private institutes. Although university programs provide some practical courses, they are not sufficient to meet both students' and broadcasting managers' needs. This environment in the Korean broadcasting industry leads to an observation that broadcasting curricula in the Korean university is more likely to focus on theories than practices.

The literature reviewed suggests that broadcasting curricula across nations is composed of theoretical, practical, or theoretical-practical programs. Besides, it is expected that technological developments and institutional changes will continue to influence the formation of broadcasting curricula from nation to nation.

Broadcasting Curricula in Three Countries

Each of the three countries studied have their own unique approach to broadcasting programs. Curricula in the United States has been described as seeking "Neo Professionalism" (Blanchard & Christ, 1993) to adjust to a newly developing media environment. The conservative educational tradition of Britain also appears ready to give way to a renovated broadcast education in response to changes in the mass media environment. Korea is experiencing tension between the existing media and the advent of new media. Universities have been at the forefront of many of those changes in Korea.

Broadcasting Curricula in the United States

Ever since the first program of journalism education at Washington College in 1869 (Sutton, 1968), curricula dealing with the mass media in the United States has experienced rapid change. The number of undergraduate programs offering a broadcasting major in the United States in 1956 numbered 86 (Sterling & Sapolsky, 1994). By 1996 it had increased to 424 (Becker & Kosicki, 1997). This internal and external growth has led to a reexamination of broadcasting curricula in the United States (Blanchard & Christ, 1993).

For example, the Syracuse experiment conducted for the model of American curricula concludes, with caution, that there exists an emerging need for the convergence of curricula between liberal arts education and professional learning (Marsh, 1988). On the other hand, some proponents for convergence favor the integration of curricula in the form of interdisciplinary studies. Specifically, the integration of theoretical and practical studies gave rise to the issue of "Neo Professionalism" (Blanchard & Christ, 1993). Neo professionalism requires a practical education combined with a strong liberal arts background.

In this regard, AEJMC (1996) concludes that the purpose of broadcast education must include "integrating liberal arts courses . . . with courses relating to the role of the mass media in society" (p. 106). This trend in the United States emphasizes integration and convergence of theoretical and practical courses.

Broadcasting Curricula in Britain

As Crook (1991), the editor of *Journalism Educator,* observed, a clear distinction between vocational education and theoretical education seems to be gradually breaking down in many European countries. However, these researchers found that the development of a combination of professional and theoretical studies into one college or university has not yet taken place at most institutions. For example, traditionally, training programs for broadcasting in Britain were basically considered "trade school" (Bannerman, 1967, p. 245). Practical training schools do not offer academic courses in their programs, whereas universities are responsible for theoretical education in the broadcasting major.

However, there are indications that some of the foundations of broadcasting curricula in Britain are beginning to change. "With the growth of cultural studies and then media studies, some more practical-oriented courses are now offered at British universities" (Morrison, 1997, p. 29). According to the most recent classifications by Orton (1997), media studies in Britain are fragmented to meet a variety of needs of learners. For those wishing to develop and obtain only skills, one- and two-year colleges and media service centers offer part-time and full-time practical courses. The Business and Technician Education Council (BTEC), Scottish Vocational Educational Council (SCOTVEC), City and Guild of London Institute (CGLI), General National Vocational Qualifications (GNVQs), General Scottish Vocational Qualifications (GSVQs), and Access Courses (AC) fall into this category and provide a pathway into practical education. Meanwhile, for those pursuing their own academic goals, three- and four-year colleges and universities offer both theoretical and practical courses. This study focuses on these institutions of higher education.

Broadcasting Curricula in Korea

Broadcast education in Korea, which began in 1954, one year after the truce of the Korean War, now encompasses more than 10,000 students at 55 universities and colleges. A growing need for university-educated broadcasters in Korea is likely to create an increasing number of students in the near future. Before 1954, formal university programs in the broadcasting major did not exist in Korea, although short courses had periodically been offered in journalism. In 1954, Hongik University began offering its first mass communication courses to students. Although Hongik University was closed due to a shortage of students later on, many institutions in the Seoul metropolitan area have offered broadcast courses. These institutions include Ewha Women's University in 1960, Hanyang University in 1963, Korea University in 1965, and Sungkyunkwan University in the school of law and political science in 1967. It is interesting to note that during this period, the department's name was in confusion. Some universities used the official title of "Department of Journalism," but a majority of schools used "Department of Communication" or "Department of Communication Arts" (Cha, 1987).

Today, 55 colleges and universities teach the broadcasting major, 15 of which offer M.A. and Ph.D. programs. Universities place emphasis on either theoretical background or practical training in their curriculum. For example, Seoul National University offers more liberal arts studies, whereas Sogang and Yonsei place emphasis more on practical competence (Heuvel & Dennis, 1993).

There are institutions established by broadcasting firms and private owners. They mainly offer practical programs including production, writing, announcing, and broadcasting technology. Except for Choongang and Sogang University, which run the practical training institutes, most of what exists outside the universities are in-house institutes of media companies such as MBC (Munwha Broadcasting Corporation) and SBS (Seoul Broadcasting System). In this regard, broadcasting education in Korea, at the university level, is still driven by the liberal arts and theoretical studies.

Research Questions

The objective of this study is to explore the current structure of broadcasting curricula and the extent to which it is being transformed by developments in the media industry. Based on the research discussed above, several questions guide this study. These questions are based on the premise that the broadcasting curricula of each country consists of courses in theory, practice, and theory-practice. Based on the brief review of these three countries, the primary research questions are:

1. What is the present status of broadcasting curricula in the United States, Britain, and Korea?
2. What is the relationship between broadcasting curricula and changes in the media environments of these countries?

Method

This study employed a content analysis to examine broadcasting curricula in the United States, Britain, and Korea. The purpose is to investigate how broadcasting curricula is constituted and to examine its relationship to the structure of media industries. To achieve this goal, three- and four-year educational institutions and undergraduate programs were analyzed.[1]

The United States Sample

The sample of the United States was based on the list of BEA institutional members who registered for 1998 membership. The list of institutions was acquired from the 1998 Winter edition of *Journal of Broadcasting & Electronic Media.* From this list, a systematic sampling of the members was used. Every third school was selected for this study. The sampling produced 100 institutions. For the analysis of specific variables, school catalogs selected in the list were collected. For uniformity and comparison with the other two countries, only the information from their catalogs was analyzed.

With an exhaustive sampling procedure, impartial chances were given to each school. The location of the schools comprises 11 (11%) western institutions, 36 (36%) midwestern institutions, 36 (36%) southern institutions, and 17 (17%) northwestern institutions. There were 68 (68%) public schools, which were funded by state governments and 32 (32%) private institutions, which were run by students' tuition and other sources of funding. Degrees offered by the institutions sampled were simply a B.A. and B.S. A total of 69 percent of the sample awarded a B.A. and 21% offered both a B.A. and a B.S. No schools offered only a B.S. degree.

British Sample

Media Courses U.K. (Orton, 1997) was used as the main analysis material. For supplemental information, the author sent e-mail messages to 98 British institutions to request the materials of a communication major. Responses were used for this study.

The book *Media Courses U.K.* consists of four parts: Courses in Further Education, Undergraduate Courses, Postgraduate Courses, and Short Courses. Further Education refers to prevocational and vocational courses; Undergraduate Courses consist of full-time colleges and universities; Postgraduate Courses award the courses for M.A., M.Phil, Ph.D. degrees; and Short Courses refer to various practical programs ranging from one-day events, through part-time evening classes, to intensive courses for several weeks.

This study utilized Part II (Undergraduate Courses) at the university level. To be consistent, junior college curriculum was excluded. This study focuses on the university level, which offers both theoretical and practical components. A total of 129 universities were drawn from the selection. The information provides degree, department of major, the amounts of practical work offered, and major course lists. The location of the universities sampled consisted of 18 (14%) London area, 6 (4.7%) Sheffield area, 7 (5.4%) Devon area, 6 (4.7%) Manchester area, 5 (3.9%) Newcastle area, 4 (3.1%) Birmingham area, 4 (3.1%) Brighton area, 3 (2.3%) Preston area, 4 (3.1%) Coventry area, 3 (2.3%) Dervy area, and 77 others.[2] Degrees awarded at British institutions were as varied as the location of the schools. Eighty-five schools (65.9%) offered B.A. (Hons), 19 (14.7%) B.A., 7 (5.4%) B.A. (Combined Hons), 7 (5.4%) B.Sc., 6 (4.7%) B.Sc. (Hons), 2 Certificates, 1 B.T.E.C. Higher National Diploma, and 1 B.Mus. (Hons).

Korean Sample

The sample of the Korean broadcast curriculum list was gleaned from *Korean Newspaper and Broadcasting Yearbook* (1997). A total of 37 institutions were listed in the book, and all 37 catalogs were compiled for this study.

Even though communication major-related departments such as the department of Public Relations and Advertising (8 departments), department of Publishing and Advertising (1), and department of Information and Broadcasting (1) are listed, those were eliminated from this study because they do not provide students with broadcast courses within majors.

The Korean Newspaper and Broadcast Yearbook included the number of enrolled students in undergraduate, master's, and Ph.D. courses, the number of faculty, their major, required major, elective major, and universal courses in major. The catalog of each university included course title, credit hours, total hours, and brief course descriptions.

Many of the institutions were located in Seoul (12 or 32.4%) and Kyungsangdo (10 or 27%). Other locations were eight (21.6%) in Chunlado, four (10.8%) in Chungcheongdo, two (5.4%) in Kyunggido, and one (2.7%) in Kangwondo. Each institution in the Korean university sample offered only a B.A. degree without a B.S. degree. In sum, a total of 266 universities and four-year colleges were content analyzed (100 from the United States, 129 from Britain, and 37 from Korea).

Units of Analysis

Since the variables available from the materials for all three countries varied, only the majors offered by departments of communication and required courses of the major were analyzed. Different terms used in the same meaning in three countries such as "core" in the United States, "mandatory modules" in Britain, and "required major" in Korea were grouped into the same category.[3]

Content analyses for curricula were conducted. The curricula classified were compared and contrasted to each other and interpreted. The classification between theoretical and practical courses in this study is significant because it is used as the standard of comparison in the broadcasting curricula in three countries.

"Theoretical Courses" refers to the courses that offer such topics as theory, history, and philosophy (McCall, 1987). "Practical courses" refers to the courses that teach students skills for practical use (e. g., production, editing, reporting, and writing). For the objectivity of analysis, reliability tests were checked by two trained graduate students. The percentage of agreement (or coefficient of reliability) for the quantitative analysis was measured. The result of measuring reliability agreement was 97.3%.[4]

Findings

The analysis of broadcasting curricula in the three countries reflected their varying interests and emphases on specific academic fields. A content analysis of 266 institutions in three countries substantiated many of the research expectations.

The first research question, which asked about the construction of the broadcasting curricula in the three countries, was partly answered by content analysis in tables. For comparison, the data collected from the United States institutions were analyzed. Table 13.1 summarizes some findings from this analysis. The most widely offered major programs in the United States institutions were Broadcasting (17, or 17%), followed by Radio/Television (11, or 11%), Mass Communication (8%), Radio/Television/Film (7%), Broadcast Journalism (6%), Communication Studies (3%), and Other (28%). As was expected, these majors offered some practical aspects in some portions of the courses, including a Television/Radio major, which offered history, theory, writing, law and ethics, and production. But these programs focused more on theoretical education than on writing or production. In contrast to Britain, the United States did not focus on Film Studies. Instead, much attention was paid to Telecommunication Studies. With the development of new media technology, the increasing interests in new visual media such as Computer, WebTV, and Satellite Broadcasting are thought to cause the increasing number of telecommunication majors.

The "Other" category was comprised of Speech Communication, Theater Arts, Telecommunication and Film, Telecommunication Arts, Media Arts and Design, Journalism and Radio-Television, Public Communication, Journalism and Broadcasting, and Broadcast Speech Communication-Dance-and-Theater.

The broadcasting curricula in Britain at the university level demonstrates that the cultural and educational tradition influenced the formation of curricula. As shown in Table 13.2, the majors offered were Television/Film/Radio (29 or 22.5%), followed by 26 (20.2%) Media Studies, 16 (12.4%) Communication Studies, and 8 (6.2%) Media Production. Even though the schools offered combined courses, which consisted of theoretical and practical components in one subject, they were more focused on the theoretical components. Another inference drawn from the curriculum analysis was that British institutions offered Historical Studies, Critical Theory, and Cultural Studies in a majority of the courses. This indicates that broadcast studies in Britain emphasize not only theoretical

Table 13.1
Broadcast Majors in U.S.

Titles	Frequencies	Percentage
Broadcasting	17	17%
Radio/Television	11	11%
Mass Communication	8	8%
Telecommunication	8	8%
Radio/Television/Film	7	7%
Broadcast Journalism	6	6%
Communication	3	3%
Communication Studies	3	3%
Journalism	3	3%
Telecommunication and Film	3	3%
Broadcast Production	3	3%
Other	28	28%
Total	100	100%

Table 13.2
Broadcast Majors in British Institutions

Titles	Frequencies	Percentage
Television/Film/Radio	29	22.5%
Media Studies	26	20.2%
Communication Studies	16	12.4%
Media Production	8	6.2%
Modern Language	5	3.9%
Media Technology	4	3.1%
Drama	4	3.1%
Mass Communication	4	3.1%
Multimedia	4	3.1%
Journalism	4	3.1%
Photography and Media	2	1.6%
Broadcast Journalism	1	0.8%
Other	18	14.0%
Total	129	97.1%[a]

[a] Percentage does not equal 100 due to rounding.

components but cultural studies. In particular, British institutions more heavily stressed Film Studies such as French Film, Italian Film, European Film History, and Film Languages than other courses. The number of Modern Language majors (5 or 3.9%) partially reflects this fact. By learning languages of other countries, students can be familiar with the media in that country. Moreover, Film Studies was linked to the study of television, video, and so-called visual media. British institutions combined Film with Television Studies in their majors.

The "Other" categories include Animation, Engineering, Computer Visualization, Imaging, Time-Based Media, Media Management, Fine Arts, Visual Communication, Humanities and Social Science, Music and Sound Recording, and Contemporary Arts.

Majors offered at the universities in Korea were not specifically classified (Table 13.3). Most of the schools named Department of Communication or Department of Mass Communication offered mass communication majors, then awarded a B.A. degree. In addition, the Korean universities offered various courses placing emphases on broadcast courses. As shown in Table 13.3, Mass Communication English courses were substantially emphasized (63 or 10.9%). Another noteworthy finding was that most departments offered History courses (90 or 15.6%) like History of Broadcasting, Broadcast History in Korea, Western History of Broadcasting and Mass Communication, and History of Communication Thoughts.

Table 13.3
Broadcast Majors in Korean Institutions

Titles	Frequencies	Percentage
History of Broadcasting	90	15.6%
Mass Communication English	63	10.9%
Mass Communication Theory	60	10.4%
Broadcasting	47	8.1%
New Media Technologies	39	6.8%
Television Production	32	5.5%
Mass Media Law	30	5.2%
Journalism	30	5.2%
Introduction to Mass Communication	25	4.3%
Film and Theater	25	4.3%
Television Programming	23	4.0%
Other	114	19.7%
Total	553[a]	100%

[a] Does not equal to 37 due to overlap.

The "Other" categories varied, consisting of Mass Media and Society, Radio Production, Effects of Mass Communication, Media Ethics, Educational Broadcasting, Mass Communication Sociology, Mass Communication Management, Topics in Mass Communication, Mass Media Economics, Mass Media in North Korea, Local Community and Mass Media, Mass Communication and Social Change, and Comparative Mass Communication.

The second research question, "What is the relationship between broadcasting curricula and changes in the media environments of these countries?" was answered by investigating course format in the three countries. The format of broadcasting curricula in the United States is somewhat different from that of Britain (Table 13.4). First, the ratio of theoretical to practical sectors is balanced and the stress on multimedia is increasing. As computer technology develops, the application of existing media theory to new media is in progress. For example, Internet Programming, Internet Design, Desktop Publishing and Editing, OnLine Research, Distance-Learning Skill, and Telecommunication Policy courses are offered as core courses in the United States institutions.

As Table 13.5 indicates, the general course format in mandatory modules in British institutions provides a map of the most common modules of the broadcast curriculum in Britain Despite some discrepancies, Communication Studies, Film Studies, and Cultural Studies are the most common mandatory courses in British institutions.

Table 13.4
Typical Course Format in Cores in the United States Institutions

Courses	Credits
Media Studies	2
Communication Studies	2
Cultural Studies	2
Film Studies	2
Television Studies	2
Media Production	2

Table 13.5
Typical Course Format in Mandatory Modules in British Institutions

Courses	Credits
Mass Communication and Society	3
News Writing and Reporting	3
Introduction to Broadcasting Techniques	3
Introduction to Communication Theory	3
Media Literacy	3
Television and Radio Production	3
Communication Law	3

Most Korean universities focused on history courses in the broadcast major (Table 13.6). Such courses as History of Korean Broadcasting, History of Western Media, or History of Communication Philosophy were the most commonly offered required courses in the broadcasting curriculum in Korean universities.

Table 13.6
Typical Course Format in Requirements in the Korean Institutions

Courses	Credits
Introduction to Mass Communication	3
History of Korean Journalism	3
Principles of Broadcasting	3
Communication Theories	3
Introduction to Journalism	3
Mass Communication Research Method	3
Mass Communication and Society	3

Conclusions and Discussion

This study was designed to compare and contrast the formation of broadcasting curricula among three countries. Especially, the extent to which educational institutions emphasized either theoretical or practical courses was compared.

Within the clear limitations of content analysis, the broadcasting curricula in the three countries offered more theoretical-based courses than practical-based ones. The findings reveal that technological developments in the media industry are the main determinants influencing the broadcasting curricula. Evidently, these curricula are being impacted by changes in the media industries. The current structure of broadcasting curricula in the three countries is reflected in their respective emphases on Telecommunications and Professional Internships in the United States, Cultural Studies and Film Studies in Britain, and History, Mass Communication, and English courses in Korea.

Findings from the analysis of U.S. institutions indicate that, among these three countries, the United States maintains the most well-balanced ratio of theory to practice in the curriculum. In view of recent trends such as an emphasis on both professionalism and a strong liberal arts background (Neo Professionalism), broadcasting curricula at U.S. institutions can be seen as responsive to industry needs.

However, it seems that more fragmentation than integration in course constitution exists in the United States broadcast curriculum. Primarily due to technological development, the number of new departments derived from communication studies is increasing. This tendency resulted in the increase and segmentation of telecommunication-related courses such as New Media and Society, Telecommunication Policy, Internet Programming, New Media Effects, and Computer Editing and Publishing.

Although the broadcasting curriculum at U.S. institutions is well organized at the macro level, certain weaknesses in the structure of curricula exist at the micro level. Despite its obvious importance, less attention was paid to Media Law and History Courses. As McCall (1987) addresses, in order for students to "assess the impact of the news, understand the implications of the First Amendment and confront ethical decisions" (p. 20), learning Media Ethics and Law is critical to broadcast students. In addition, a History of World Media as well as American Media should be offered to students. Only one-fifth of the four-year institutions in the United States offered a history course in the broadcasting major. As Kittross (1989) pointed out, history courses in institutions have been ignored, so that students are taught something superficial. Therefore, the research suggests that more history courses need to be offered.

The analysis of broadcasting curricula at British institutions illustrated that they primarily stressed Film and Cultural Studies. This suggests that the philosophical, social, and cultural climate has influenced the broadcasting curriculum in Britain. Another noteworthy result was that most of the visual communications such as television, film, computer, and multimedia were offered as "combined courses" in one category. In turn, most broadcast courses offered theory and practices in the same course. These characteristics of broadcasting majors at British institutions reflect a combination of the European educational tradition and the more recently emerged computerized and electronic methods of communication. These institutions emphasize the cultural background of broadcasting as well as the utility of new visual technology.

While the basic emphases remain on Communication Theory, Media Studies, and Cultural Studies, a remarkable expansion in Television, Film, Radio, and Multimedia Production courses was observed from this analysis. Broadcasting curricula at British institutions can be said to be integrating into "theoretical practices" (French & Richard, 1994).

Another characteristic of broadcast curricula at British institutions is the convergence between the needs of the media industries and the skills and knowledge provided by the courses. In other words, the format of the recent broadcasting curricula in Britain is reflected by an increased emphasis on the application of knowledge. A large number of combined courses in broadcast majors are an example. About one-third of the course content consisting of practice, which is based on students' theoretical background, was taught by professors in class.

While it is apparent that liberal arts and theoretical courses are taught in most British institutions, the importance of several courses seems to be neglected in the curriculum. As Theory, Critical, and Cultural Studies are significant, so are Law, Policy, Management, and Effects of Mass Communication. These courses have been defined by Webster (1989) as "Administrative Studies." To provide the most stimulating learning environment, a combination of critical studies and administrative studies in media education should be secured (Webster, 1989). Currently, broadcasting curricula at British institutions lack this administrative component, which also includes Media Economics and Policy classes.

For practical education in the broadcasting major in Britain, an internship program is recommended. Although most British institutions offered practical classes, internships were not common. Internships can be said to occupy a position "at the center of a schism between teachers of theory and skill courses," and they seem to be more necessary than ever before in the information era, which requires learners to adapt to new technology (Ciofalo, 1989).

In the past, learning the practical components in mass media was not as hard a process as it has since become. The recent survey done by Medsger (1996) suggests that mass media educators and journalists should understand and use new and diverse technologies. Medsger warns, however, that this takes time. Still, the utilization of media technologies in the broadcasting curriculum has become a critical component of higher education. Further, interns who encounter new technologies firsthand in the workplace also need to be prepared. In this area, broadcasting curricula in the United States institutions generally offered well-prepared internship programs.

The imbalance between theoretical and practical education was rather apparent in the analysis of the broadcasting curriculum of Korean institutions. A heavy emphasis on History and English courses was found at all universities in Korea. Although professional education is offered by private institutions, additional practical components at the major universities seem to be necessary.

The most emphasized course at the Korean universities was History. Almost all universities offered history courses as a requirement in the major, especially a History of Korean Mass Communication. In order to deepen the educational experience (Kittross, 1989), universities require students to develop the ability to criticize and contemplate the development of mass communication. This emphasis on history courses is consistent with the goal of liberal arts and theory-based education, which aims at the development of critical insights and analytical thinking.

In addition, the theoretical emphasis in the broadcasting curriculum in Korean universities is a result of the recruitment system in the media industry (Heuvel & Dannis, 1993). The examination for entering a broadcasting company requires applicants to be proficient in Language (English), History, Literature, Politics, in turn, and have a broad liberal arts background. This system causes broadcasting students to concentrate on theoretical studies more so than practical training. Accordingly, although students receive practical training such as program production and news reporting at private institutes, a well-balanced education at the university level is lacking. Another implication from the findings is that more telecommunication courses are needed. Such majors as Computer Mediated Communication, New Technology, or Telecommunication Policy should be offered.

As broadcast facilities continue to expand, television and radio production courses can be developed on Korean campuses in response to private institutes' production programs. Also, professional internship programs can provide students with practical experience and academic credit. Since attending private practical training schools is costly and lacks a liberal arts education, offering practical courses at universities would meet the goal of university education, which is "to produce well-rounded graduates who have critical-thinking skills as well as practical skills" (AEJMC, 1996, p. 102). The ideas learned in theories can be directly tied to creating programs when combined with broadcasters' intuitions. Program ideas do not spring from blank slates. Well-educated professionals are more likely to have ideas that will be successfully developed as media products.

Noticeably, the authors found that there is no standard in "core" ("mandatory modules" in Britain, and "required" in Korea) and "elective" (free-choice) courses in all three countries sampled. Thus, drawing a unified course map was almost impossible. For example, although there were preferred core courses, some core courses at one university were included as elective courses at another university or vice versa in three countries. Accordingly, more standardization of course modules would make it easier to draw conclusions. Although it is impossible to produce unified course standardization in elective courses, core (required) course standardization can provide a map of a country's own structure of broadcasting curricula.

The broadcasting courses offered in all three countries are comprised of mostly combined or mixed content in both lectures and practices. However, single production courses were also considered important required or elective courses. Theory and practice examination are both important because these results can provide a clue as to how to construct a better format for a broadcasting curriculum. As Duhe and Zukowski (1997) suggest, the best type of broadcast curriculum benefits students, programs, and the media industry. In this view, deriving a model from the examination of the three countries' broadcasting curricula can be helpful for educators and students as they prepare for their future.

As media technology and new systems continue to develop, the efforts to weave communication studies into a body of knowledge that includes education in practical skills have been made by most educational institutions (Iorio & Williamson, 1995). An updated version of this trend was also found throughout this study in the form of growing interest by institutions in multimedia technologies. The development of a broadcasting curriculum is allowing for more interdisciplinary and combined courses.

CONCLUSION

Coming back to the original question of this study, the results found that each country had its own unique characteristics when it comes to designing a broadcasting curriculum. At the same time, all three countries are experiencing some restructuring of their curricula in response to technological developments in the media industry. The United States institutions are well balanced in both theoretical and practical courses. The United States institutions are more likely to respond in a timely manner to technological developments such as multimedia. British institutions are driven more by theoretical courses. However, even these universities are shifting some emphasis toward the inclusion of practical skills. In Korea, the educational orientation in broadcasting is still focused on theoretical knowledge. Students learn practical skills at private institutes at the "vocational school" level.

Taken together, broadcasting curricula in the three countries are influenced by technological developments and a changing media landscape. The interplay of these forces with higher education can be said to be a global trend.

NOTES

1. A bachelor's degree in the European educational system including Britain comprises three years (Morrison 1997). The model of Russian media and journalism education is changing (*Journalism and Mass Communication Educator, 52 (3),* 26–34). And there are four-year colleges at the United States university level. Therefore, this study examines three-year and four-year colleges and universities.
2. Other areas include Cambridge, Bath, Pool Dorset, West Yorkshire, Middlesex, Twickenham, Buckingham, Leicester, Bedford, Essex, Norwich, Glamorgan, Glasgow, Lancaster, Leeds, Hull, Lincoln, Luton, Edinburgh, Nottingham, Milton Keynes, Oxford, Ayr, Paisley, Portsmouth, and Hants area.
3. "Core" is defined as "the form of prescribed courses or sets of courses within a fundamental area" (Blanchard & Christ 1985. "In search of the unit core." Commonalities in curricula. *Journalism Educator, 40,* 28–33). According to their classification, for example, universal core consists of utilitarian, introduction, and integrative. "Module" is defined as "basic building block of the course." Modules consist of mandatory, elective, and free choice modules (BA [honors] information and media studies, 1997. *Definitive Course Document and Student Handbook 1997–1998, 1,* [Brochure]. London: University of Brighton). "Major" in Korea refers to the courses offered in department. It includes required and elective courses.
4. The formula is:

$$\% = \frac{2M}{N1 + N2}$$

   Where M is the number of coding decisions on which the two judges are in agreement and N1 and N2 indicate the number of coding decisions made by judges 1 and 2, respectively (Holsti 1969).

REFERENCES

Association for Education in Journalism and Mass Communication. (1996). A report on the findings of the AEJMC curriculum task force. *Journalism & Mass Communication Educator, 50 (4),* 102–119.

Bannerman, L. (1967). Broadcast training in Britain. *Journalism and Mass Communication Educator, 11* (3), 245–251.

Barner, D. L. (1996). A national sample of recent graduates address broadcast curriculum. *Feedback, 37* (2), 22–25.

Becker, L. B., & Kosicki, G. M. (1997). Annual survey of enrollment and degrees awarded. *Journalism and Mass Communication Educator, 52 (3),* 63–74.

Blanchard, R. O., & Christ, W. G. (1993). *Media education and the liberal arts: A blueprint for the new professionalism.* Hillsdale, NJ: Lawrence Erlbaum Associates.

Broadcast Education Association Constitution (1990). Washington, DC: Author.

Cha, B. G. (1987). *Introduction to mass communication.* Seoul: Seyongsa.

Chang, T. (1997). The world of journalism and mass communication education. *Journalism and Mass Communication Educator, 52 (3),* 4–7.

Ciofalo, A. (1989). Legitimacy of internships for academic credit remains controversial. *Journalism Educator, 44 (4),* 25–31.

Crook, J. A. (1991). European from training association with help from European community. *Journalism Educator, 45 (4),* 92–96.

Duhe, S. F., & Zukowsli, L. A. (1997). Radio-TV journalism curriculum: First jobs and career preparation. *Journalism and Mass Communication Educator, 52 (1),* 4–15.

French, D., & Richards, M. (1994). Theory, practice and market forces in Britain: A case of relative autonomy. In D. French & M. Richards (Eds.), *Media education across Europe.* London, New York: Routledge.

Giglioli, P. P. (1994). Italy: The coming of age of media studies. In D. French & M. Richards (Eds.), *Media education across Europe.* London, New York: Routledge.

Heuvel, J. V., & Dennis, E. E. (1993). The unfolding lotus: East Asia's changing media. New York, NY: The Freedom Forum Media Studies Center.

Holsti, O. R. (1969). *Content analysis for the social sciences and humanities.* Massachusetts: Addison-Wesley.

Iorio, S. H., & Williamson, K. (1995). The role of liberal arts courses within communication curricula. *Journalism Educator, 50 (1),* 16–25.

Kim, K., Kim, W. Y., & Kang, J. G. (1994). *Broadcasting in Korea.* Seoul: Nanam.

Kittross, J. K. (1989, Fall). Six decades of education for broadcasting . . . and counting. *Feedback, 39 (4),* 30–41.

Marsh, P. T. (1988). The Syracuse experiment. In P. T. Marsh (Ed.), *Contesting the boundaries of liberal and professional education: The Syracuse experiment.* (pp. 9–33). New York: Syracuse University Press.

McCall, J. M. (1987). Liberal arts focus provides training for media careers. *Journalism Educator, 42,* 17–21.

Medsger, B. (1996). *Winds of change: Challenges confronting journalism education.* Arlington, Virginia: Freedomforum.

Monush, B. (1995). *International television and video almanac.* (40th ed.). New York, NY: Auiqley Publishing Company.

Morrison, J. (1997). The changing model of Russian media and journalism education. *Journalism and Mass Communication Educator, 52 (3),* 26-34.

Murray, M. D. (1992). The basic course in mass communication. In M. D. Murray and A. J. Ferry (Eds.), *Teaching mass communication: A guide to better instruction.* New York, NY: Praeger.

Niven, H. (1961). The development of broadcasting education in institutions of higher education. *Journal of Broadcasting, 5,* 241–250.

Orton, L. (1997). *Media courses U.K.* (4th Ed.). London: British Film Institute.

Pennybacker, J. H. (1965). Working with universities. *Journal of Broadcasting, 9 (2),* 183–187.

Renz, B. B. (1991). Yes to announcing. *Feedback, 32 (2),* 22–23.

Sterling, C., & Sapolsky, B. S. (Eds.) (1994). Trends and changes in broadcast education. *BEA Directory of Media Programs in North American Universities and Colleges*, 8–9.

Sutton, A. A. (1968). *Education for journalism in the United States from its beginning to 1940.* New York, NY: AMS Press.

Webster, J. (1989). Media study in a time of technological change. *Feedback, 30 (3)*, 20–23.

Conclusion

In the early years of the 21st century, broadcasting programs at colleges and universities are facing critical challenges concerning how best to prepare students for media careers of the future. As the chapters in this work demonstrate, many of these challenges are the result of changes in technology and revolve around the need for academic programs to make the commitment to accompany the broadcast industry's transition from analog to digital equipment. Doing so is not an inexpensive process, and there may well be institutions that have neither the resources nor the student interest to make such a commitment. But those with strong broadcasting programs cannot afford to resist such a transition. Clearly, programs that make the decision to upgrade from analog to the new industry standard of digital equipment will emerge as leaders in the field of broadcast education. Programs that fail to do so appear destined to be relegated to secondary status.

Perhaps the most critical issue facing universities and broadcasting programs will be how to fund such a large-scale transition. In most cases the answer is likely to involve a gradual phase-in of new equipment over a period of years. Broadcasters are often taking this approach themselves, so the strategy is certainly a sensible one, especially considering that industry standards for new production and transmission equipment can be expected to establish themselves over time. One of the authors addressed the finance issue by describing a model in which designated student fees are utilized for projects such as this one. We understand that increasing student fees is not a popular issue to raise on college campuses these days, either with students, parents, administrators, or in the political arena of state politics. However, the example at North Texas strikes us as an especially appropriate one because the students who pay the fees are also the ones who benefit directly and immeasurably from the availability of state-of-the-art equipment in the major that they have chosen to pursue. That situation is quite different from many fees that are levied on all students but often fund areas that not all take an interest in. At the very least, the model merits consideration as administrators and faculty contemplate the future of their academic broadcasting programs.

At the same time, the questions for broadcast educators go far beyond simply making sure that the latest in equipment and technology is available for their students. Deeper issues involve the impact of new technology on effective teaching styles and course content, the development of entirely new emphases for curricula, the importance of interdisciplinary collaboration, and the need to prepare broadcast journalism students for careers that include reporting across media platforms as well as traditional radio and television storytelling. All these challenges exist amid an academic context that must include a commitment to the traditional pedagogical values in higher education, such as factual information gathering, the development of a clear and compelling writing style, and an emphasis on the development of critical thinking and skills of analysis and interpretation. It has been our goal from the beginning of this project to create a forum for the thoughts of some of the top broadcast educators in the country and by doing so to provide a valuable resource, perhaps even a road map for faculty and administrators to refer to when making important decisions about curricula and equipment.

As such, we conclude by addressing each of the matters just mentioned and offering suggestions for broadcast educators to pursue as they revise and refine their programs in light of changes in the media industry.

TEACHING DIGITAL MEDIA

Several new issues have been identified when it comes to teaching broadcast courses involving the application of digital equipment. First, in many cases students come to these classes, especially production classes, with the expectation that learning new software will enhance their career prospects and the likelihood that they will qualify for well-paying jobs after graduation. There are several implications of such an expectation. The first and most obvious is that instructors need to do all they can to prepare themselves to help students learn the new software. Most of the time doing so will require professional development support from the department or university in the form of workshops or seminars. Because of continual changes and updates in software, professional development in this area will become an ongoing priority for faculty. At the same time, there will always be some students who are even more efficient at navigating the software than the instructors who teach the class. This situation refers back to the fact that, in most cases, children and teenagers are growing up in a world that is immensely more computer literate than the one that raised the parents and teachers who mentor them. Although this can be an uncomfortable reality for the instructor, it need not be an embarrassment or a major obstacle to the teaching and learning process. At least in the short term, instructors should cultivate students who have demonstrated expertise as lab assistants or graduate assistants who can be valuable resources for the other students in the class. The important consideration is that in the long term, instructors must become as committed to their own lifelong learning as they are encouraging their students to become.

Along with an emphasis on ongoing professional development in the area of software skills, it should be emphasized that knowing the software is only one part of the educational process. Even a university instructor with little expertise in this area will still have a greater depth of knowledge, maturity, and professional experience at understanding what constitutes an effective media product. Is the story convincing and well told? Is the communication clear and effective? How skillfully is the communication message crafted? What does it look like and feel like to the consumer? What response will the message elicit in the audience? In the end it is the content and the quality of the media product that matters far more than the tools used to create it.

Developing a quality media product requires the ability to think through the project from beginning to end. Students must continually revise and edit until unnecessary elements have been pared away and what remains constitutes exactly what the media practitioner envisioned to communicate the message. Further, although software skills are important, it has been emphasized in these chapters that it is unlikely students will be utilizing the exact same software by the time they graduate. Updates make it common that a newer version or a different software altogether will have become the standard when the graduate begins a professional career. Soon that version in turn will be supplanted by a newer one. What typically matters is not expertise in a particular software but rather the ability to understand and to navigate a system, to learn new systems as they are developed, and, in a broader intellectual sense, to adapt to change. Along these lines, what will ultimately matter most to a student's career is not the technological wizardry that might help land them that first job but rather the qualities inherent in a truly educated person—inquisitive, self-challenging, and ever improving. It is these qualities that will enable graduates to qualify for positions of increasing responsibility, such as administration and management, as they mature and develop throughout a lifetime. As part of their role in higher education, broadcast educators must remind students and sometimes themselves that their first calling is not strictly vocational but instead cultivating in students a love of learning, a devotion to quality and performing at one's best, and an appreciation of the challenge and the beauty of the life of the mind.

Finally, the impact of digital media on teaching and learning styles needs to be addressed. The role of the instructor in the modern age is best perceived as one of a coach who works alongside the student and who understands the student as an individual with strengths and areas for improvement. The coaching model is one that broadcast and print educators have long embraced as a way to help writers and reporters develop, but it has never been more important than now in the digital age. The instructor is far more than a lecturer or someone who imparts information, and the student is far more than a mere recipient of knowledge and facts. The process of education has become infinitely more interactive to the point that at times lines between teacher and student almost seem to blur. The relationship should be highly positive, mutually energizing, individual, and unique. This relationship might be better understood as one of collaboration, partnership, or something similar to that between an expert and a client or a coach and a valuable member of the team who has something important to contribute.

THE EDUCATOR AS COLLABORATOR

An additional point made apparent by this study is that successful broadcast educators in the digital age will collaborate with more than their students. Of course, broadcast educators from various colleges and universities have always collaborated with each other at academic conferences, seminars, and other events. Now, however, the lines of contact have been extended well beyond the discipline. Converging technologies have created a situation in which disciplines that previously had little if anything in common now have much to be gained by combining expertise. Departments of music, art, and computer science, to name a few, have all emerged as areas where broadcasters may look to develop innovative interdisciplinary programs that can offer the foundation of careers in areas such as animation, audio recording, and digital media.

At the same time, chairs and others who lead broadcasting departments must be careful that these efforts are not misconstrued by administrators as indicating that the traditional broadcasting major is somehow less relevant than it was before. This point is a crucial one, because it is at the administrative level where decisions will be made about the allocation of resources and faculty lines. The message to communicate is that traditional areas of specialization remain vibrant while new areas of expertise are emerging due to digital technology. Interdisciplinary approaches provide a meaningful corollary to emphases in production, news gathering and reporting, management, and advertising that have long been the cornerstones of a strong broadcasting program and will continue to be so in the future. The objective should be to attract additional students to the department who might otherwise have majored in a different discipline or attended another college or university.

Of course, it is imperative that department leaders do more than merely state the case verbally. A commitment must be made to engage in detailed fact gathering. The current data on student majors, credit hours, and full-time enrollment must be examined carefully and updated continually to recognize and track trends across periods of years. This way the argument in favor of the expanding broadcasting major will always be an honest one that is based on verifiable facts.

The chapter on multimedia reporting demonstrates how collaboration between the previously distinct areas of print journalism and broadcast journalism can better prepare students for the reporting careers of a converging news industry. Academic programs across the country have been responding in a wide variety of ways to the issue of media convergence. Some have chosen a completely new path and begun offering only a convergence emphasis. Others have remained more traditional and kept the boundaries distinct. Still many others have begun to chart a course somewhere in the middle. In the final analysis, the correct decision depends on the needs and capabilities—financial and otherwise—of each academic program, as well as the values of the institution and the needs of students being served. It seems likely to us, however, that some classroom experience in convergence, combined with in-depth knowledge of the unique qualities of radio, television, print, and the Internet, will distinguish the exceptional broadcasting programs from the rest of the pack.

The term "convergence" refers largely to the common digital technological base that has emerged to produce and deliver the media product and in some, but not all, cases to receive it. Students will need some basic knowledge of how to package a story differently for multiple platforms. But in a larger context, audiences as consumers of media products will always have very different uses for radio, television, print, and the Internet. Detailed understanding of one or more of these media provides a student with stronger preparation and a greater chance for advancement in a media career. In addition, there are numerous entry-level opportunities in areas such as production, news reporting, on-air program host, promotion, and advertising that will remain specific to a particular medium.

THE FUTURE

A number of priorities stand out as broadcast faculty plan for the future. The first priority is quite simple really: remain true to the core values of higher education. A thorough understanding of key principles, insights, and ethics will always be paramount. Strong writing skills will always be vitally important to a person's success, perhaps more so now than ever. The attraction of new technology creates a temptation to become mesmerized by what remains at bottom merely a way of getting one's message across. The quality of the message, not just the sophistication of the tool, lies at the heart of effective communication. The acquisition of appropriate skills, the application of analytical thought, the ability to conceive of a project with a clear beginning, middle, and ending, and the mastery of effective storytelling techniques all remain imperative cornerstones for a university-educated broadcasting major.

Beyond these core values, educators would be well advised to heed the advice of the experts whose views find expression in this volume. The impact of digital technology has already proven pervasive, and discussion of the implications of this change must become a central part of the curriculum. The media is changing rapidly toward an emphasis on viewer control and interactivity. At the same time that an appreciation of topics such as the history of the media remains important, additional skill sets pertaining to the Internet will become an important part of a broadcast graduate's repertoire. So will an understanding of how the Web can best be utilized to promote broadcasting and other entertainment outlets, in particular in the larger context of learning about the business nature of the industry. Further, an awareness of the global marketplace has become increasingly critical, especially in the areas of program syndication and distribution.

Most educators believe that, for the immediate future, it will be important to continue providing some training on analog equipment, at the very least so that students appreciate the technological changes that have occurred. On a practical level, it will prepare them for employment at those media outlets that are still in the process of transition. Other priorities will simply never phase out, especially in broadcast journalism, where ethics, validating sources and facts, and the emphasis on local coverage and public service remain crucial.

The early 21st century is an exciting time to be in the profession of broadcast education. Like their industry counterparts in radio and television throughout the decades, broadcasting faculty are proving themselves to be innovative, enthusiastic, and adaptable when confronted with rapid technological change. Many are already engaged in important research about digital technology. In the years ahead, such research will no doubt go far beyond the emphasis of this study on new technology and the design of academic programs. It will address the global issues of how we as broadcast educators, media practitioners, students, and citizens can become communicators who more effectively utilize technology to understand, interact with, and improve our world.

Index

About the Editors and Contributors

Joseph R. Blaney is Assistant Professor of Communication and Director of Broadcasting at Illinois State University. His interests are in image repair and maintenance.

Mark J. Braun is Associate Dean at Gustavus Adolphus College. His interests include gender and media law and policy.

Dom Caristi is Associate Professor of Telecommunications at Ball State University. His interests are in media law and interactive media.

Joyce Chen is Associate Professor of Communication Studies at the University of Northern Iowa. Her interests are in media technology and information.

Jerry Condra is Assistant Professor of Communication at the State University of New York at Oswego. His interests are in broadcast management and sales.

Steve Craig is Professor of Radio-TV-Film at the University of North Texas. His interests include media criticism and history.

Gerard Donnelly is Associate Professor and Chair of Mass Communication at Northwest Missouri State University. His interests include media leadership and radio broadcasting.

W.A. Kelly Huff is Visiting Professor of Journalism and Mass Communication at the University of Georgia. His interests include radio, broadcast journalism, and media law and regulation.

Matt Jenkins is Associate Professor of Communication at Cameron University. His interests include video production and broadcast technologies.

Jong G. Kang is Professor of Communication at Illinois State University. His interests are in broadcast management and media effects.

Seok Kang is Assistant Professor of Communication at Arkansas Technical University.

Kevin C. Lee is Assistant Professor of Communication at Western Carolina University. His interests are in media ethics and pedagogy.

Rebecca Ann Lind is Associate Professor of Communication and Assistant Vice Chancellor for Research at the University of Illinois-Chicago. Her interests include gender, radio, and broadcast journalism.

Bruce Mims is Associate Professor of Mass Communication and Assistant Vice Chancellor for Research at Southeast Missouri State University. His interests include radio, new technologies, and media law and regulation.

Rick Musser is Professor in the William Allen White School of Journalism at the University of Kansas. His interests are in print journalism and media convergence.

Matt Rouch is Instructor of Mass Communication at Northwest Missouri State University. His interests include television production and scriptwriting.

Doug Sudhoff is Assistant Professor of Mass Communication at Northwest Missouri State University. His interests are in broadcast news and television production.